Springer Topics in Signal Processing

Volume 15

More information about this series at http://www.springer.com/series/8109

Orhan Gazi

Polar Codes

A Non-Trivial Approach to Channel Coding

Springer

Orhan Gazi
Department of Electronic and
Communication Engineering
Çankaya University
Etimesgut/Ankara
Turkey

ISSN 1866-2609 ISSN 1866-2617 (electronic)
Springer Topics in Signal Processing
ISBN 978-981-13-0736-2 ISBN 978-981-13-0737-9 (eBook)
https://doi.org/10.1007/978-981-13-0737-9

Library of Congress Control Number: 2018943249

Printed on acid-free paper

This Springer imprint is published by the registered company Springer Nature Singapore Pte Ltd.
part of Springer Nature
The registered company address is: 152 Beach Road, #21-01/04 Gateway East, Singapore 189721, Singapore

Contents

Chapter 1
Information Theory Perspective of Polar Codes and Polar Encoding

Polar codes are one of the recently discovered capacity achieving channel codes. What makes the polar codes different from other channel codes is that polar codes are designed mathematically and their performance are mathematically proven. On the other hand, the previous channel codes are constructed in a trivial manner. In this chapter we first provide the fundamental concepts from the information theory, such as entropy, mutual information, and channel capacity, then discuss the philosophy behind the idea of polar encoding considering the fundamentals subjects of the information theory.

After explaining the philosophy of the polar encoding, we inspect the polar encoding technique prosed by Arıkan [1]. Polar encoding operation can be performed in a recursive manner. For this purpose, there are two types of encoder structures defined in [1]. After explaining the recursive encoding methods, we provide the mathematical formulas for the polar encoding operation.

1.1 Information Theory Perspective of Polar Codes

For the discrete random variable $\tilde{U}$, the entropy $H\left(\tilde{U}\right)$ is defined as

$$H\left(\tilde{U}\right) = -\sum_{u} p(u) \log p(u) \tag{1.1}$$

where $p(u)$ is the probability mass function of $H\left(\tilde{U}\right)$, and $\log(\cdot)$ is the logarithmic function with base 2. The joint entropy of two discrete random variables $\tilde{U}$ and $\tilde{Y}$ is defined as

$$H\left(\tilde{U}, \tilde{Y}\right) = -\sum_{u,y} p(u, y) \log p(u, y) \tag{1.2}$$

O. Gazi, *Polar Codes*, Springer Topics in Signal Processing 15,
https://doi.org/10.1007/978-981-13-0737-9_1

where $p(u, y)$ is the joint probability mass function of $\tilde{U}$ and $\tilde{Y}$.

The mutual information between two discrete random variable $\tilde{U}$ and $\tilde{Y}$ is defined as

$$I\left(\tilde{U};\tilde{Y}\right) = H\left(\tilde{U}\right) - H\left(\tilde{U}\middle|\tilde{Y}\right) \text{or as } I\left(\tilde{U};\tilde{Y}\right) = H\left(\tilde{Y}\right) - H\left(\tilde{Y}\middle|\tilde{U}\right) \quad (1.3)$$

which can be expressed in terms of the joint and marginal probability mass functions as

$$I\left(\tilde{U};\tilde{Y}\right) = \sum_{u,y} p(u, y) \log \frac{p(u, y)}{p(u)p(y)}. \quad (1.4)$$

The expression in (1.4) is nothing but a probabilistic average quantity and can be written in short as

$$I\left(\tilde{U};\tilde{Y}\right) = E\left\{\log \frac{p\left(\tilde{U},\tilde{Y}\right)}{p\left(\tilde{U}\right)p\left(\tilde{Y}\right)}\right\}. \quad (1.5)$$

Note:

$$E\left(g\left(\tilde{U},\tilde{Y}\right)\right) = \sum_{u,y} p(u, y)g(u, y).$$

The mutual information between $\tilde{U}$ and $(\tilde{Y}, \tilde{Z})$ is defined as

$$I\left(\tilde{U};\tilde{Y},\tilde{Z}\right) = \sum_{u,y,z} p(u, y, z) \log \frac{p(u, y, z)}{p(u)p(y, z)} \quad (1.6)$$

which can be written as

$$I\left(\tilde{U};\tilde{Y},\tilde{Z}\right) = E\left\{\log \frac{p\left(\tilde{U},\tilde{Y},\tilde{Z}\right)}{p\left(\tilde{U}\right)p\left(\tilde{Y},\tilde{Z}\right)}\right\}. \quad (1.7)$$

In a similar manner, we can define $I\left(\tilde{U}_1, \tilde{U}_1\tilde{Y}_1, \tilde{Y}_2\right)$ as

$$I\left(\tilde{U}_1, \tilde{U}_2;\tilde{Y}_1, \tilde{Y}_2\right) = \sum_{u_1,u_2,y_1,y_2} p(u_1, u_2, y_1, y_2) \log \frac{p(u_1, u_2, y_1, y_2)}{p(u_1, u_2)p(y_1, y_2)} \quad (1.8)$$

In general, for two random variable vectors

$$\bar{U} = \left[\tilde{U}_1, \tilde{U}_2, \ldots \tilde{U}_N\right] \quad Y = \left[\tilde{Y}_1, \tilde{Y}_2, \ldots \tilde{Y}_N\right] \quad (1.9)$$

the mutual information $I(\bar{U};\bar{Y})$ is defined as

$$I(\bar{U};\bar{Y}) = \sum_{\bar{u},\bar{y}} p(\bar{u},\bar{y}) \log \frac{p(\bar{u},\bar{y})}{p(\bar{u})p(\bar{y})}. \tag{1.10}$$

Theorem 1.1 *The mutual information*

$$I\left(\tilde{U};\tilde{Y},\tilde{Z}\right)$$

can be written as

$$I\left(\tilde{U};\tilde{Y},\tilde{Z}\right) = I\left(\tilde{U};\tilde{Y}\right) + I\left(\tilde{U};\tilde{Z}\middle|\tilde{Y}\right). \tag{1.11}$$

Proof 1.1 From definition (1.5), we can write $I\left(\tilde{U};\tilde{Y},\tilde{Z}\right)$ as

$$I\left(\tilde{U};\tilde{Y},\tilde{Z}\right) = \sum_{u,y,z} p(u,y,z) \log \frac{p(u,y,z)}{p(u)p(y,z)} \tag{1.12}$$

where

$$\frac{p(u,y,z)}{p(u)p(y,z)}$$

can be written as

$$\frac{p(u,y,z)}{p(u)p(y,z)} = \frac{p(u,z|y)p(y)}{p(u)p(z|y)p(y)}$$

in which multiplying the numerator and denominator of the right side by $p(u,y)$ we get

$$\frac{p(u,y,z)}{p(u)p(y,z)} = \frac{p(u,y)}{p(u,y)} \frac{p(u,z|y)p(y)}{p(u)p(z|y)p(y)}$$

where the right hand side can be rearranged as

$$\frac{p(u,y)}{p(u)p(y)} \frac{p(u,z|y)p(y)}{p(z|y)p(u,y)}$$

in which using $p(u,y) = p(u|y)p(y)$, we get

$$\frac{p(u,y)}{p(u)p(y)} \frac{p(u,z|y)p(y)}{p(z|y)p(u|y)p(y)}$$

which is simplified as

$$\frac{p(u,y)}{p(u)p(y)}\frac{p(u,z|y)}{p(z|y)p(u|y)}.$$

Thus, we have shown that

$$\frac{p(u,y,z)}{p(u)p(y,z)}=\frac{p(u,y)}{p(u)p(y)}\frac{p(u,z|y)}{p(z|y)p(u|y)} \tag{1.13}$$

Using (1.13) in (1.12), we obtain

$$I\left(\tilde{U};\tilde{Y},\tilde{Z}\right)=\sum_{u,y,z}p(u,y,z)\log\frac{p(u,y)}{p(u)p(y)}\frac{p(u,z|y)}{p(z|y)p(u|y)} \tag{1.14}$$

which can be written as

$$I\left(\tilde{U};\tilde{Y},\tilde{Z}\right)=\underbrace{\sum_{u,y}p(u,y)\log\frac{p(u,y)}{p(u)p(y)}}_{I(\tilde{U};\tilde{Y})}+\underbrace{\sum_{u,y,z}p(u,y,z)\log\frac{p(u,z|y)}{p(z|y)p(u|y)}}_{I\left(\tilde{U};\tilde{Z}\middle|\tilde{Y}\right)}$$

leading to

$$I\left(\tilde{U};\tilde{Y},\tilde{Z}\right)=I\left(\tilde{U};\tilde{Y}\right)+I\left(\tilde{U};\tilde{Z}\middle|\tilde{Y}\right).$$

Example 1.1 Show that

$$I\left(\tilde{U}_1,\tilde{U}_2;\tilde{Y}_1,\tilde{Y}_2\right)=I\left(\tilde{U}_1;\tilde{Y}_1,\tilde{Y}_2\right)+I\left(\tilde{U}_2,\tilde{Y}_1;\tilde{Y}_2\middle|\tilde{U}_1\right)$$

Solution 1.1 Considering the probability expressions in the logarithmic parts of the terms in

$$I\left(\tilde{U}_1,\tilde{U}_2;\tilde{Y}_1,\tilde{Y}_2\right)=I\left(\tilde{U}_1;\tilde{Y}_1,\tilde{Y}_2\right)+I\left(\tilde{U}_2;\tilde{Y}_1,\tilde{Y}_2\middle|\tilde{U}_1\right)$$

we can make the matching

$$I\left(\tilde{U}_1,\tilde{U}_2;\tilde{Y}_1,\tilde{Y}_2\right)\rightarrow\frac{p(u_1,u_2,y_1,y_2)}{p(u_1,u_2)p(y_1,y_2)}$$

$$I\left(\tilde{U}_1;\tilde{Y}_1,\tilde{Y}_2\right)\rightarrow\frac{p(u_1,y_1,y_2)}{p(u)p(y_1,y_2)}$$

$$I\left(\tilde{U}_2;\tilde{Y}_1,\tilde{Y}_2\middle|\tilde{U}_1\right)\rightarrow\frac{p(u_2,y_1,y_2|u_1)}{p(u_2|u_1)p(y_1,y_2|u_1)}.$$

And, it can be shown that

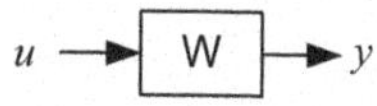

Fig. 1.1 Discrete memoryless channel

$$\frac{p(u_1, u_2, y_1, y_2)}{p(u_1, u_2)p(y_1, y_2)} = \frac{p(u_1, y_1, y_2)}{p(u_1)p(y_1, y_2)} \times \frac{p(u_2, y_1, y_2|u_1)}{p(u_2|u_1)p(y_1, y_2|u_1)}$$

which means that

$$I\left(\tilde{U}_1, \tilde{U}_2; \tilde{Y}_1, \tilde{Y}_2\right) = I\left(\tilde{U}_1; \tilde{Y}_1, \tilde{Y}_2\right) + I\left(\tilde{U}_2; \tilde{Y}_1, \tilde{Y}_2 \middle| \tilde{U}_1\right).$$

Theorem 1.2 *For two random variable vectors*

$$\bar{U} = \left[\tilde{U}_1, \tilde{U}_2, \ldots, \tilde{U}_N\right] \quad \bar{Y} = \left[\tilde{Y}_1, \tilde{Y}_2, \ldots, \tilde{Y}_N\right]$$

the mutual information $I\left(\bar{U}; \bar{Y}\right)$ *can be written as*

$$I(\bar{U}; \bar{Y}) = \sum_{i=1}^{N} I\left(\tilde{U}_i; \bar{Y} \middle| \tilde{U}_1, \tilde{U}_2, \ldots, \tilde{U}_{i-1}\right). \tag{1.15}$$

Example 1.2 The mutual information

$$I\left(\tilde{U}_1, \tilde{U}_2, \tilde{U}_3, \tilde{U}_4; \tilde{Y}_1, \tilde{Y}_2, \tilde{Y}_3, \tilde{Y}_4\right)$$

can be written as

$$\begin{aligned} I\left(\tilde{U}_1, \tilde{U}_2, \tilde{U}_3, \tilde{U}_4; \tilde{Y}_1, \tilde{Y}_2, \tilde{Y}_3, \tilde{Y}_4\right) = {} & I\left(\tilde{U}_1; \tilde{Y}_1, \tilde{Y}_2, \tilde{Y}_3, \tilde{Y}_4\right) + I\left(\tilde{U}_2; \tilde{Y}_1, \tilde{Y}_2, \tilde{Y}_3, \tilde{Y}_4 \middle| \tilde{U}_1\right) \\ & + I\left(\tilde{U}_3; \tilde{Y}_1, \tilde{Y}_2, \tilde{Y}_3, \tilde{Y}_4 \middle| \tilde{U}_1, \tilde{U}_2\right) \\ & + I\left(\tilde{U}_4; \tilde{Y}_1, \tilde{Y}_2, \tilde{Y}_3, \tilde{Y}_4 \middle| \tilde{U}_1, \tilde{U}_2, \tilde{U}_3\right). \end{aligned}$$

1.1.1 The Philosophy of Polar Codes

Consider the discrete memoryless channel shown in Fig. 1.1 where it is seen that the symbol u is passed through the discrete memoryless channel indicated by the symbol W.

The symbol at the input of the discrete memoryless channel can be considered as generated by a discrete random variable $\tilde{U}$, similarly the symbol at the output of the channel can be considered as generated by another discrete random variable $\tilde{Y}$. Then, the discrete memoryless channel of Fig. 1.1 can be shown as in Fig. 1.2.

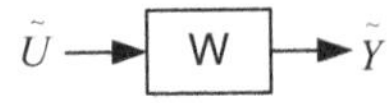

Fig. 1.2 Discrete memoryless channel with random variables

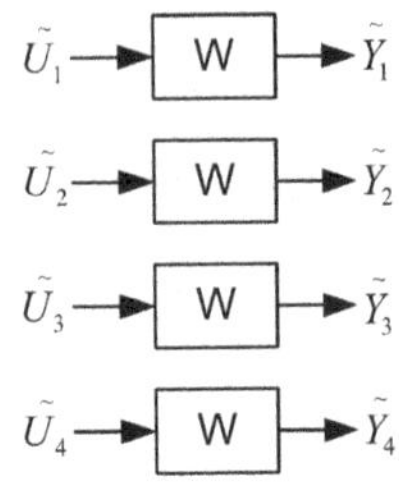

Fig. 1.3 Parallel discrete memoryless channels

Consider the transmission of 4 different symbols $[u_1 u_2 u_3 u_4]$ through the channel in a serial manner. In this case, we can assume that these 4 symbols are generated by *IID* random variables, and we can think the transmission of each symbol through the channel separately as in Fig. 1.3.

Let $\bar{U} = \left[\tilde{U}_1\, \tilde{U}_2\, \tilde{U}_3\, \tilde{U}_4\right]$ and $\bar{Y} = \left[\tilde{Y}_1\, \tilde{Y}_2\, \tilde{Y}_3\, \tilde{Y}_4\right]$, then the mutual information between $\bar{U}$ and $\bar{Y}$ can be written as

$$I(\bar{U};\bar{Y}) = I\left(\tilde{U}_1;\bar{Y}\right) + I\left(\tilde{U}_2;\bar{Y}|\tilde{U}_1\right) + I\left(\tilde{U}_3;\bar{Y}|\tilde{U}_1,\tilde{U}_2\right) + I\left(\tilde{U}_4;\bar{Y}|\tilde{U}_1,\tilde{U}_2,\tilde{U}_3\right) \tag{1.16}$$

If $\tilde{U}_1$ is independent of $\tilde{Y}_1, \tilde{Y}_2, \tilde{Y}_3, \tilde{Y}_4$, then we can write that

$$I\left(\tilde{U}_1;\bar{Y}\right) = I\left(\tilde{U}_1;\tilde{Y}_1\right). \tag{1.17}$$

Let's verify the correctness of (1.17). Using definition of mutual information, we can write $I\left(\tilde{U}_1;\bar{Y}\right)$ as

$$I\left(\tilde{U}_1;\bar{Y}\right) = \sum_{u_1,y_1,y_2,y_3,y_4} p(u_1,y_1,y_2,y_3,y_4)\log\frac{p(u_1,y_1,y_2,y_3,y_4)}{p(u_1)p(y_1,y_2,y_3,y_4)} \tag{1.18}$$

where due to independence, the probabilities $p(u_1,y_1,y_2,y_3,y_4)$ and $p(y_1,y_2,y_3,y_4)$ can be written as

$$p(u_1,y_1,y_2,y_3,y_4) = p(u_1,y_1)\,p(y_2)\,p(y_3)\,p(y_4) \quad p(y_1,y_2,y_3,y_4) = p(y_1)\,p(y_2)\,p(y_3)\,p(y_4).$$

Then, (1.18) reduces to

$$I\left(\tilde{U}_1;\bar{Y}\right)=\sum_{u_1,y_1,y_2,y_3,y_4} p(u_1,y_1)p(y_2)p(y_3)p(y_4)\log\frac{p(u_1,y_1)p(y_2)p(y_3)p(y_4)}{p(u_1)p(y_1)p(y_2)p(y_3)p(y_4)}$$

which can be simplified as

$$I\left(\tilde{U}_1;\bar{Y}\right)=\sum_{u_1,y_1} p(u_1,y_1)\log\frac{p(u_1,y_1)}{p(u_1)p(y_1)}$$

which is equal to $I\left(\tilde{U}_1;\tilde{Y}_1\right)$. Thus, we showed that

$$I\left(\tilde{U}_1;\bar{Y}\right)=I\left(\tilde{U}_1;\tilde{Y}_1\right).$$

In a similar manner, we can show that

$$I\left(\tilde{U}_2;\bar{Y}|\tilde{U}_1\right)=I\left(\tilde{U}_2;\tilde{Y}_2\right)$$

$$I\left(\tilde{U}_3;\bar{Y}|\tilde{U}_1,\tilde{U}_2\right)=I\left(\tilde{U}_3;\tilde{Y}_3\right)$$

$$I\left(\tilde{U}_4;\bar{Y}|\tilde{U}_1,\tilde{U}_2,\tilde{U}_3\right)=I\left(\tilde{U}_4;\tilde{Y}_4\right).$$

Then, (1.16) can be written as

$$I(\bar{U};\bar{Y})=I\left(\tilde{U}_1;\tilde{Y}_1\right)+I\left(\tilde{U}_2;\tilde{Y}_2\right)+I\left(\tilde{U}_3;\tilde{Y}_3\right)+I\left(\tilde{U}_4;\tilde{Y}_4\right). \tag{1.19}$$

Let $C=\max I(\bar{U};\bar{Y})$, then we have

$$\max I(\bar{U};\bar{Y})=\underbrace{\max I\left(\tilde{U}_1;\tilde{Y}_1\right)}_{C}+\underbrace{\max I\left(\tilde{U}_2;\tilde{Y}_2\right)}_{C}+\underbrace{\max I\left(\tilde{U}_3;\tilde{Y}_3\right)}_{C}+\underbrace{\max I\left(\tilde{U}_4;\tilde{Y}_4\right)}_{C}\rightarrow$$

$$\max I(\bar{U};\bar{Y})=4C.$$

1.1.2 Fundamental Idea of Polar Codes

Find a method that the channel outputs depend on the other inputs as well. That is,

$$I\left(\tilde{U}_1;\bar{Y}\right)\neq I\left(\tilde{U}_1;\tilde{Y}_1\right)$$

$$I\left(\tilde{U}_2;\bar{Y}|\tilde{X}_1\right) \neq I\left(\tilde{U}_2;\tilde{Y}_2\right)$$

$$I\left(\tilde{U}_3;\bar{Y}|\tilde{U}_1,\tilde{U}_2\right) \neq I\left(\tilde{U}_3;\tilde{Y}_3\right)$$

$$I\left(\tilde{U}_4;\bar{Y}|\tilde{U}_1,\tilde{U}_2,\tilde{U}_3\right) \neq I\left(\tilde{U}_4;\tilde{Y}_4\right)$$

That is, the right hand of

$$I(\bar{U};\bar{Y}) = I\left(\tilde{U}_1;\bar{Y}\right) + I\left(\tilde{U}_2;\bar{Y}|\tilde{U}_1\right) + I\left(\tilde{U}_3;\bar{Y}|\tilde{U}_1,\tilde{U}_2\right) + I\left(\tilde{U}_4;\bar{Y}|\tilde{U}_1,\tilde{U}_2,\tilde{U}_3\right) \tag{1.20}$$

cannot be simplified as in (1.19). In addition, we try to assure that

$$I\left(\tilde{U}_1;\bar{Y}\right) < I\left(\tilde{U}_3;\bar{Y}|\tilde{U}_1,\tilde{U}_2\right) \leq I\left(\tilde{U}_2;\bar{Y}|\tilde{U}_1\right) < I\left(\tilde{U}_4;\bar{Y}|\tilde{U}_1,\tilde{U}_2,\tilde{U}_3\right)$$

i.e., capacities increases in an orderly manner but the total capacity stays constant, i.e., $I(\bar{U};\bar{Y}) = 4C$. In fact, we want to assure that as the length of information sequence $N \to \infty$, we want the channel capacities to be equal either to 0 or to 1.

1.2 Polar Encoding Operation

Let N be the length of information sequence. Consider the transmission of two information bits u_1 and u_2. The simplest polar encoder structure for $N = 2$, i.e., for $\bar{u} = [u_1 u_2]$, is depicted in Fig. 1.4.

From Fig. 1.4, we can write that

$$x_1 = u_1 \oplus u_2 \quad x_2 = u_2.$$

For the information word $\bar{u} = [u_1\, u_2]$, after polar encoding operation, the obtained code-word is $\bar{x} = [x_1\, x_2]$ where $x_1 = u_1 \oplus u_2$ and $x_2 = u_2$. The relation between $\bar{x}$ and $\bar{u}$ for $N = 2$ can mathematically be expressed as

$$\bar{x} = \bar{u}G \tag{1.21}$$

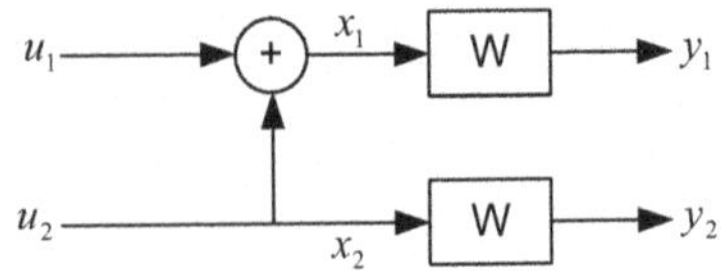

Fig. 1.4 Polar encoder for $N = 2$ with discrete memoryless channels

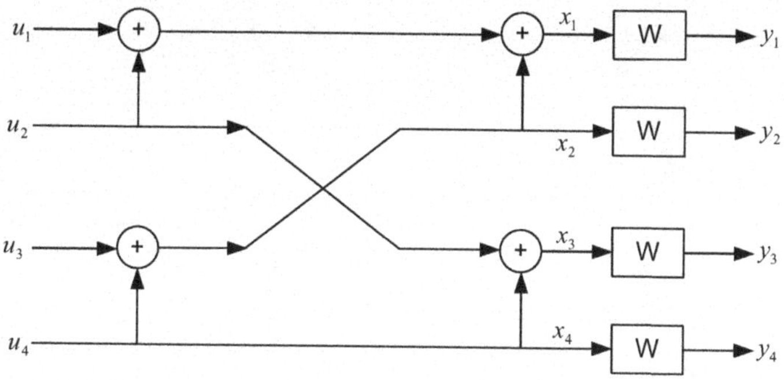

Fig. 1.5 Polar encoder for $N = 4$ with discrete memoryless channels

where G is the generator matrix for $N = 2$ and it is equal to

$$G_2 = \begin{bmatrix} 1 & 0 \\ 1 & 1 \end{bmatrix}.$$

For $N = 4$, polar encoder structure is depicted in Fig. 1.5. From Fig. 1.5, we get

$$\begin{aligned} x_1 &= u_1 \oplus u_2 \oplus u_3 \oplus u_4 \\ x_2 &= u_3 \oplus u_4 \\ x_3 &= u_2 \oplus u_4 \\ x_4 &= u_4 \end{aligned}$$

which can be written as

$$\bar{x} = \bar{u} \times G$$

where

$$\bar{x} = [x_1\, x_2\, x_3\, x_4] \quad \bar{u} = [u_1\, u_2\, u_3\, u_4]$$

and the generator matrix G equals to

$$G_4 = \begin{bmatrix} 1 & 0 & 0 & 0 \\ 1 & 0 & 1 & 0 \\ 1 & 1 & 0 & 0 \\ 1 & 1 & 1 & 1 \end{bmatrix}. \tag{1.22}$$

The encoder structure in Fig. 1.5 can be redrawn using straight lines as in Fig. 1.6.

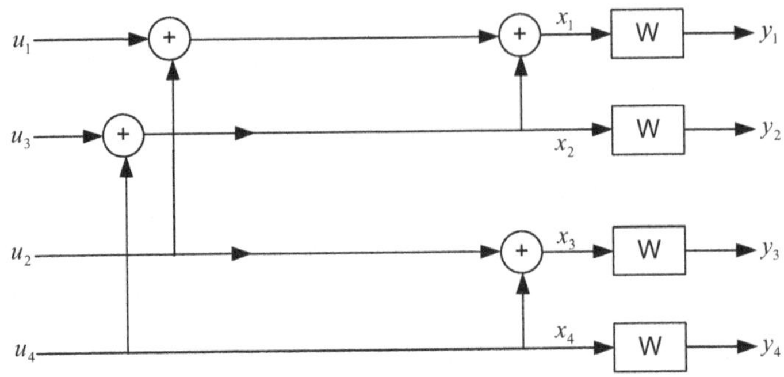

Fig. 1.6 Redrawn polar encoder for $N = 4$ with discrete memoryless channels

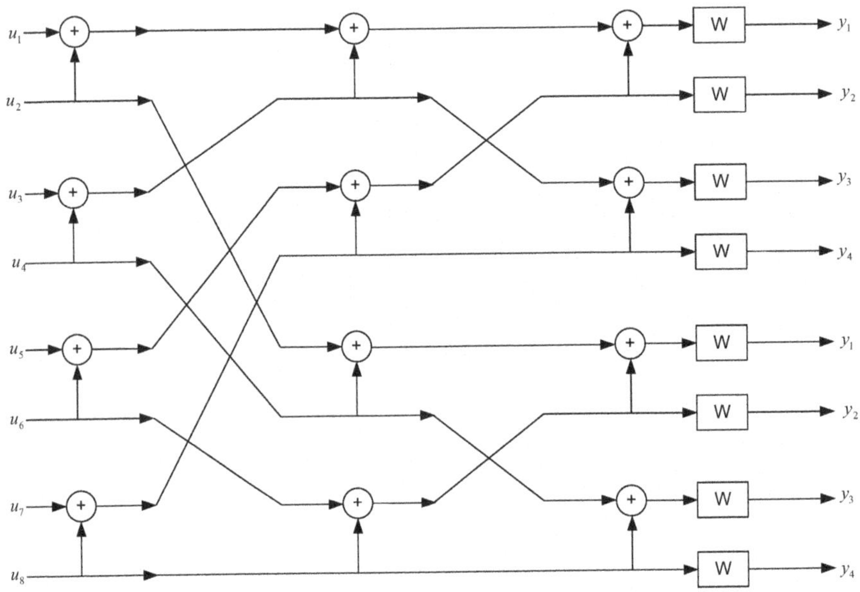

Fig. 1.7 Polar encoder for $N = 8$ with discrete memoryless channels

Although the encoder structure of Fig. 1.6 seems to be different from the one in Fig. 1.5, they are the same of each other and the relation between $\bar{u}$ and $\bar{x}$ stays the same in both structures.

Note that, in Fig. 1.5 the input sequence is $\bar{u} = [u_1\, u_2\, u_3\, u_4]$ whereas in Fig. 1.6 the input sequence is $\bar{u} = [u_1\, u_3\, u_2\, u_4]$.

For $N = 8$, polar encoder structure is depicted in Fig. 1.7.

From Fig. 1.7, we get

$$x_1 = u_1 \oplus u_2 \oplus u_3 \oplus u_4 \oplus u_5 \oplus u_6 \oplus u_7 \oplus u_8$$

$$x_2 = u_5 \oplus u_6 \oplus u_7 \oplus u_8$$
$$x_3 = u_3 \oplus u_4 \oplus u_7 \oplus u_8$$
$$x_4 = u_7 \oplus u_8$$
$$x_5 = u_2 \oplus u_4 \oplus u_6 \oplus u_8$$
$$x_6 = u_6 \oplus u_8$$
$$x_7 = u_4 \oplus u_8$$
$$x_8 = u_8$$

which can be written as

$$\bar{x} = \bar{u} \times G$$

where

$$\bar{x} = [x_1\, x_2\, x_3\, x_4\, x_5\, x_6\, x_7\, x_8] \quad \bar{u} = [u_1\, u_2\, u_3\, u_4\, u_5\, u_6\, u_7\, u_8]$$

and the generator matrix G equals to

$$G_8 = \begin{bmatrix} 1\,0\,0\,0\,0\,0\,0\,0 \\ 1\,0\,0\,0\,1\,0\,0\,0 \\ 1\,0\,1\,0\,0\,0\,0\,0 \\ 1\,0\,1\,0\,1\,0\,1\,0 \\ 1\,1\,0\,0\,0\,0\,0\,0 \\ 1\,1\,0\,0\,1\,1\,0\,0 \\ 1\,1\,1\,1\,0\,0\,0\,0 \\ 1\,1\,1\,1\,1\,1\,1\,1 \end{bmatrix} \tag{1.23}$$

The encoder structure in Fig. 1.7 can be redrawn using straight lines as in Fig. 1.8.

Although the encoder structure of Fig. 1.7 seems to be different from the one in Fig. 1.8, they are the same of each other and the relation between $\bar{u}$ and $\bar{x}$ stays the same in both structures. Note that, in Fig. 1.7 the input sequence is $\bar{u} = [u_1\, u_2\, u_3\, u_4\, u_5\, u_6\, u_7\, u_8]$ whereas in Fig. 1.6 the input sequence is $\bar{u} = [u_1\, u_5\, u_3\, u_7\, u_2\, u_6\, u_4\, u_8]$.

1.3 Recursive Construction of Polar Encoder Structures

There are two techniques introduced in [1] for the recursive construction of polar encoder structures. In this section, we will explain both techniques in details.

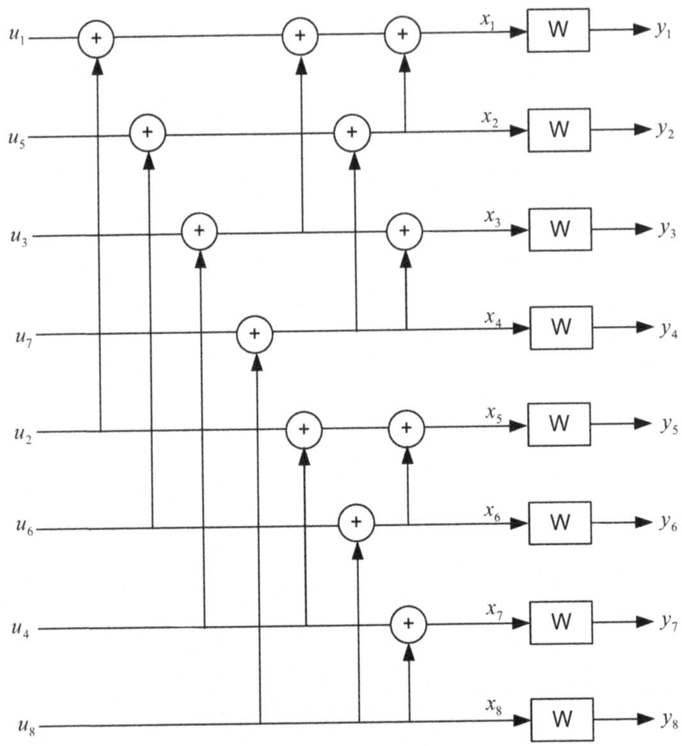

Fig. 1.8 Redrawn polar encoder for $N = 8$ with discrete memoryless channels

Fig. 1.9 W_2 channel

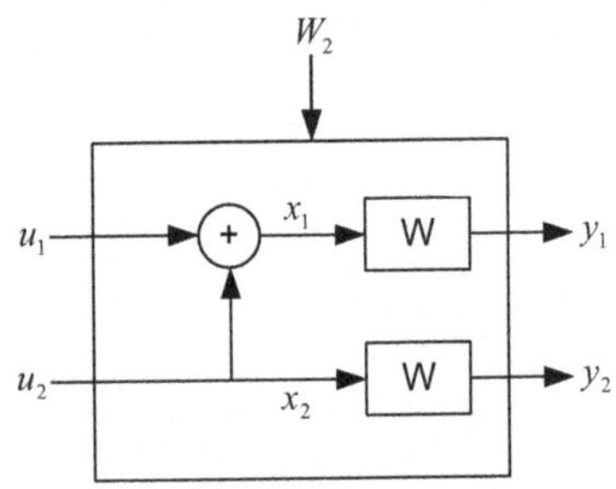

Method-1:

We will denote the encoder structure involving N information bits by W_N. For instance, the polar encoder structure for $N = 2$ shown in Fig. 1.9 is denoted by W_2.

Polar encoder structure for $N = 4$ is depicted in Fig. 1.10 where it is seen that the encoder structure W_4 consists of two W_2 structures and one permutation matrix indicated by R_4. It is seen from Fig. 1.10 that

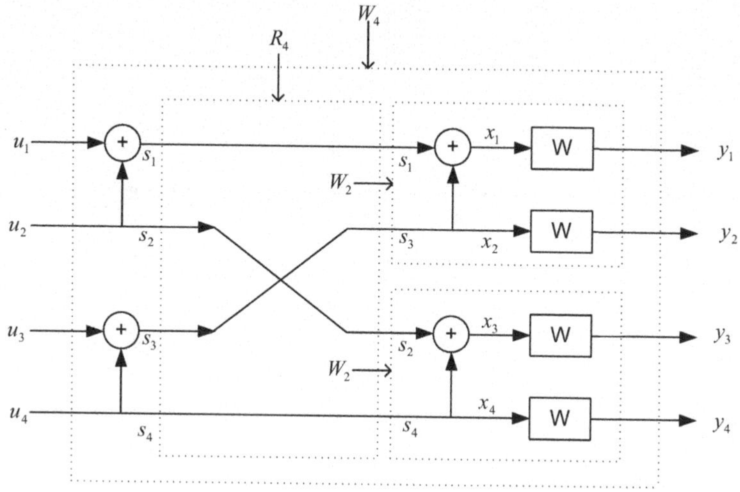

Fig. 1.10 Construction of W_4 channel from two W_2 channels

$$[s_1\, s_2\, s_3\, s_4] \times R_4 = [s_1\, s_3\, s_2\, s_4].$$

The construction of W_8 from two W_4 channels are illustrated in Fig. 1.11 where the permutation matrix R_8 satisfy

$$[s_1\, s_2\, s_3\, s_4\, s_5\, s_6\, s_7\, s_8] \times R_8 = [s_1\, s_3\, s_5\, s_7\, s_2\, s_4\, s_6\, s_8].$$

Now, it is time to consider the general structure for the construction of polar encoder structures in a recursive manner. The first method for the recursive construction of the polar encoder structures is depicted in Fig. 1.12 where R_N is the odd-even permutation matrix.

Example 1.3 Using the recursive encoder construction structure depicted in Fig. 1.12, we can construct the encoder structure for $N = 4$, i.e., the structure W_4 using two W_2 and one R_4 as in (Fig. 1.13).

Example 1.4 Using the recursive encoder construction structure depicted in Fig. 1.12, we can construct the encoder structure for $N = 8$, i.e., the structure W_8 using two W_4 and one R_8 as in Fig. 1.14.

Method-2:

An alternative recursive polar encoder structure equivalent to that of the one in Fig. 1.12 is depicted in Fig. 1.15 where R_N is the odd-even permutation matrix.

Example 1.5 Using the recursive structure in Fig. 1.15, the polar encoder for $N = 4$ is constructed as in Fig. 1.16.

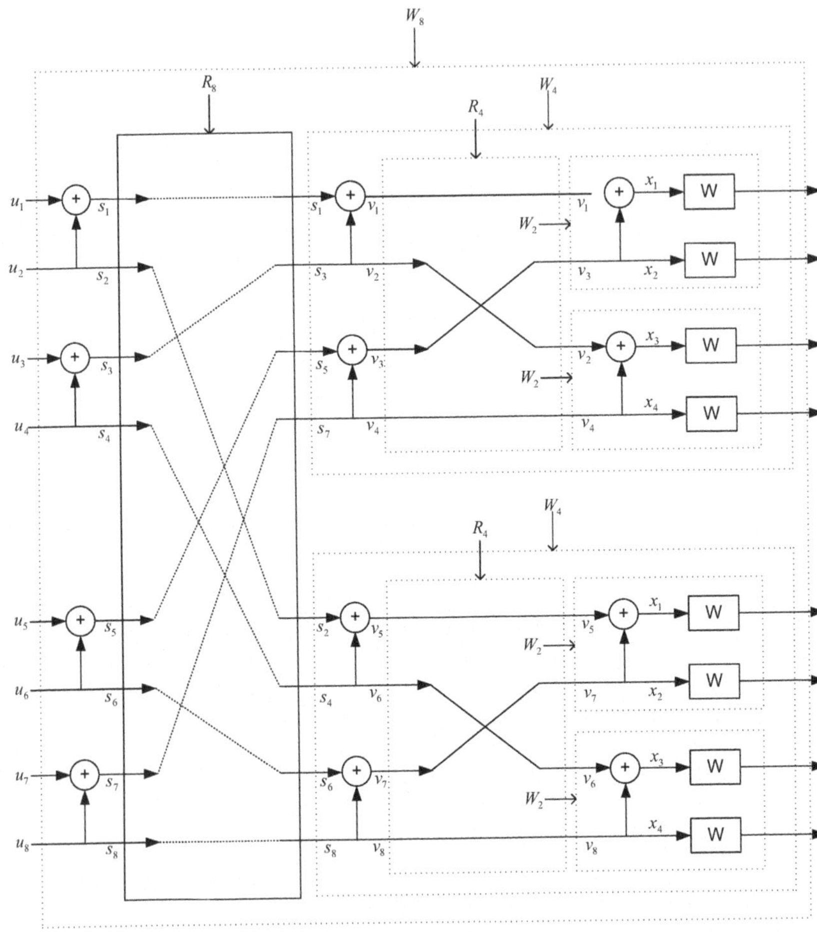

Fig. 1.11 Construction of W_8 channel from two W_4 channels

Example 1.6 Using the recursive structure in Fig. 1.15, the polar encoder for $N = 8$ is constructed as in Fig. 1.17.

Using straight lines, the encoder structure in Fig. 1.17 can be redrawn as in Fig. 1.8.

1.4 Generator Matrix and Encoding Formula for Polar Codes

In this section, we will provide information about the construction of generator matrix of the polar codes. Let A, and B be two matrices, the Kronecker product of these two matrices is defined as:

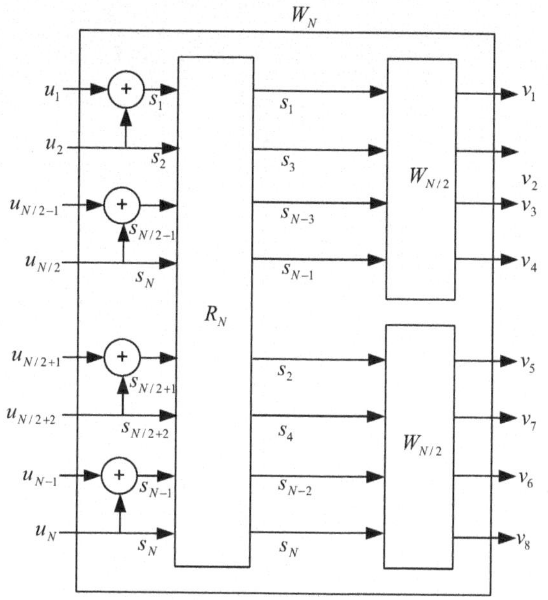

Fig. 1.12 Polar encoder general structure [1]

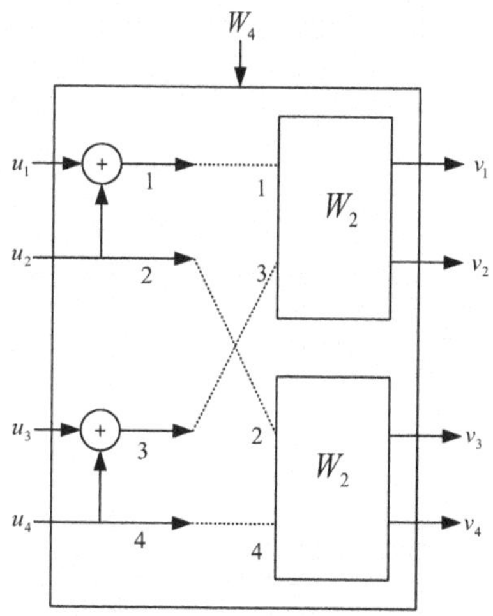

Fig. 1.13 The polar encoder for $N = 4$

$$A \otimes B = \begin{bmatrix} a_{1,1}B \ldots a_{1,N}B \\ \ddots \\ a_{N,1}B \ldots a_{N,N}B \end{bmatrix}. \tag{1.24}$$

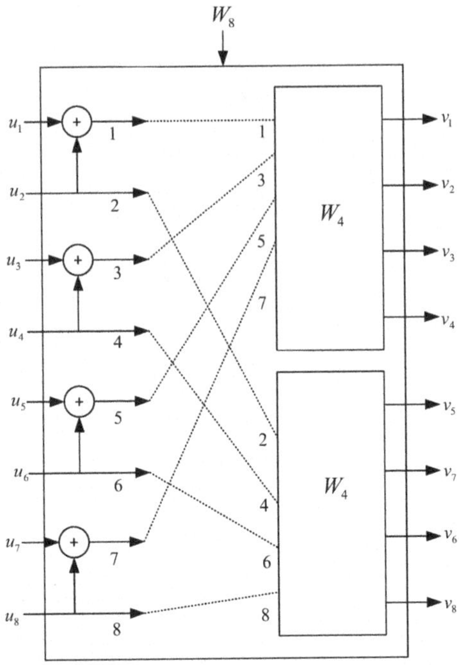

Fig. 1.14 The polar encoder for $N = 8$

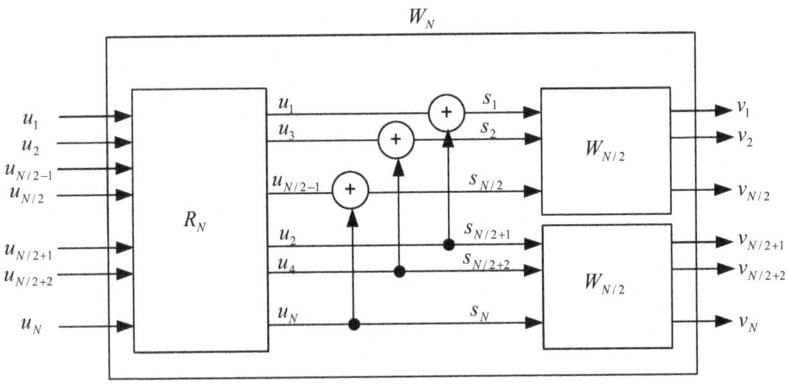

Fig. 1.15 Alternative structure for general polar encoder [1]

The data bit vector u_1^N is defined as

$$u_1^N = (u_1, u_2, \ldots, u_N). \tag{1.25}$$

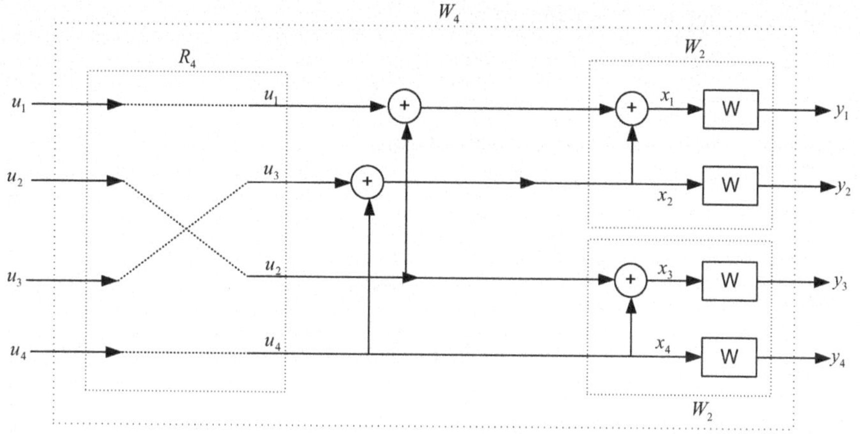

Fig. 1.16 Alternative polar encoder for $N = 4$

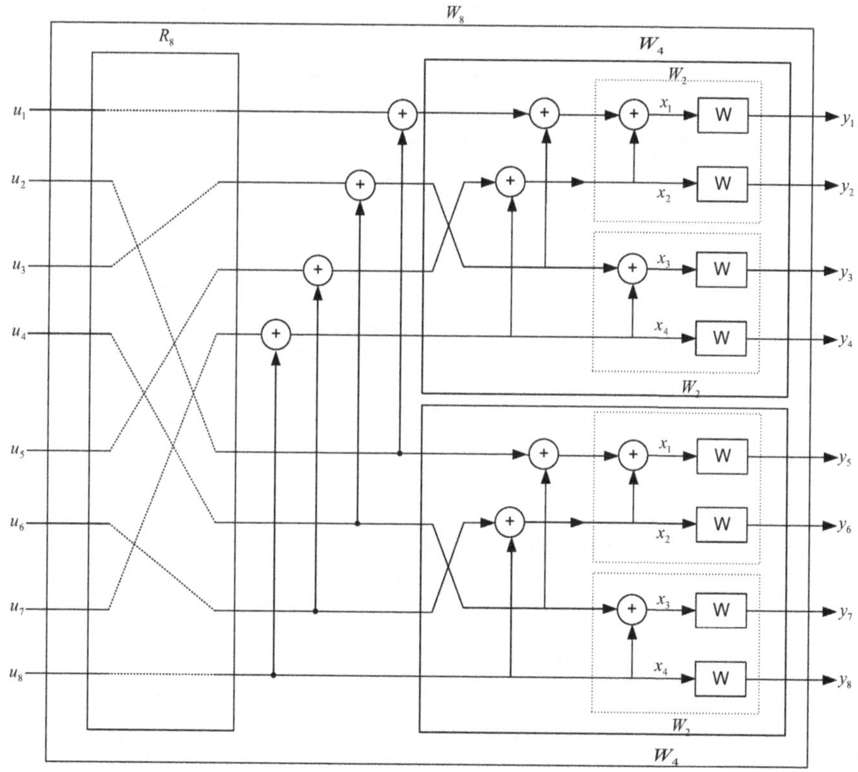

Fig. 1.17 Alternative polar encoder for $N = 8$

Polar encoding of u_1^N is achieved via

$$x_1^N = u_1^N G_N \tag{1.26}$$

where the generator matric G_N is obtained using

$$G_N = B_N F^{\otimes n} \tag{1.27}$$

in which $N = 2^n$, $F = \begin{bmatrix} 1 & 0 \\ 1 & 1 \end{bmatrix}$ and B_N is calculated using

$$B_N = R_N \left(I_2 \otimes B_{\frac{N}{2}} \right) \tag{1.28}$$

where the initial value of B_N, i.e., B_2 is I_2, and R_N denotes the $N \times N$ reverse shuffle permutation matrix defined by

$$(s_1, s_2, s_3, \ldots, s_N) R_N = (s_1, s_3, \ldots, s_{N-1}, s_2, s_4, \ldots, s_N). \tag{1.29}$$

For instance, for (s_1, s_2, s_3, s_4) the matrix R_4 is found from

$$(s_1, s_2, s_3, s_4) R_4 = (s_1, s_3, s_2, s_4)$$

as

$$R_4 = \begin{bmatrix} 1 & 0 & 0 & 0 \\ 0 & 0 & 1 & 0 \\ 0 & 1 & 0 & 0 \\ 0 & 0 & 0 & 1 \end{bmatrix}.$$

Since

$$(s_1, s_2, s_3, s_4) \begin{bmatrix} 1 & 0 & 0 & 0 \\ 0 & 0 & 1 & 0 \\ 0 & 1 & 0 & 0 \\ 0 & 0 & 0 & 1 \end{bmatrix} = (s_1, s_3, s_2, s_4).$$

In a similar manner, we can find the matrix R_8 as

$$R_8 = \begin{bmatrix} 1 & 0 & 0 & 0 & 0 & 0 & 0 & 0 \\ 0 & 0 & 0 & 0 & 1 & 0 & 0 & 0 \\ 0 & 1 & 0 & 0 & 0 & 0 & 0 & 0 \\ 0 & 0 & 0 & 0 & 0 & 1 & 0 & 0 \\ 0 & 0 & 1 & 0 & 0 & 0 & 0 & 0 \\ 0 & 0 & 0 & 0 & 0 & 0 & 1 & 0 \\ 0 & 0 & 0 & 1 & 0 & 0 & 0 & 0 \\ 0 & 0 & 0 & 0 & 0 & 0 & 0 & 1 \end{bmatrix}.$$

Using (1.27), we can calculate the generator matrix of the polar code for $N = 2$, $N = 4$, $N = 8$, and $N = 16$ as

$$G_2 = \begin{bmatrix} 1\,0 \\ 1\,1 \end{bmatrix} \quad G_4 = \begin{bmatrix} 1\,0\,0\,0 \\ 1\,0\,1\,0 \\ 1\,1\,0\,0 \\ 1\,1\,1\,1 \end{bmatrix} \quad G_8 = \begin{bmatrix} 1\,0\,0\,0\,0\,0\,0\,0 \\ 1\,0\,0\,0\,1\,0\,0\,0 \\ 1\,0\,1\,0\,0\,0\,0\,0 \\ 1\,0\,1\,0\,1\,0\,1\,0 \\ 1\,1\,0\,0\,0\,0\,0\,0 \\ 1\,1\,0\,0\,1\,1\,0\,0 \\ 1\,1\,1\,1\,0\,0\,0\,0 \\ 1\,1\,1\,1\,1\,1\,1\,1 \end{bmatrix}. \tag{1.30}$$

When the generator matrices G_2, G_4, and G_8 are inspected, it is seen that we can obtain G_4 from G_2 by constructing two matrices G_{2z}, G_{2r} and concatenating them column-wise as

$$G_4 = \begin{bmatrix} G_{2z} \\ G_{2r} \end{bmatrix} \tag{1.31}$$

where G_{2z} is obtained by inserting zero-column vectors between each two columns, i.e., if G_2 is indicated as

$$G_2 = [\bar{c}_1\ \bar{c}_2]$$

then G_{2z} is constructed as

$$G_{2z} = \left[\bar{c}_1\ \bar{0}\ \bar{c}_2\ \bar{0}\right]$$

and the matrix G_{2r} is constructed duplicating the columns as

$$G_{2r} = [\bar{c}_1\ \bar{c}_1\ \bar{c}_2\ \bar{c}_2].$$

In a similar manner, if G_4 is indicated as

$$G_4 = [\bar{c}_1\ \bar{c}_2\ \bar{c}_3\ \bar{c}_4]$$

then G_{4z}, G_{4r} are construed as

$$G_{4z} = \left[\bar{c}_1\,\bar{0}\,\bar{c}_2\,\bar{0}\,\bar{c}_3\,\bar{0}\,\bar{c}_4\,\bar{0}\right] \quad G_{4r} = [\bar{c}_1\ \bar{c}_1\ \bar{c}_2\ \bar{c}_2\ \bar{c}_3\ \bar{c}_3\ \bar{c}_4\ \bar{c}_4]$$

and G_8 is formed as

$$G_8 = \begin{bmatrix} G_{4z} \\ G_{4r} \end{bmatrix}.$$

For $N = 8$, if we have G_8, we can construct G_{16} as

$$G_{16} = \begin{bmatrix} G_{8z} \\ G_{8r} \end{bmatrix}.$$

The generator matrix G_{16} can be found as in

$$G_{16} = \begin{bmatrix} 1&0&0&0&0&0&0&0&0&0&0&0&0&0&0&0 \\ 1&0&0&0&0&0&0&0&1&0&0&0&0&0&0&0 \\ 1&0&0&0&1&0&0&0&0&0&0&0&0&0&0&0 \\ 1&0&0&0&1&0&0&0&1&0&0&0&1&0&0&0 \\ 1&0&1&0&0&0&0&0&0&0&0&0&0&0&0&0 \\ 1&0&1&0&0&0&0&0&1&0&1&0&0&0&0&0 \\ 1&0&1&0&1&0&1&0&0&0&0&0&0&0&0&0 \\ 1&0&1&0&1&0&1&0&1&0&1&0&1&0&1&0 \\ 1&1&0&0&0&0&0&0&0&0&0&0&0&0&0&0 \\ 1&1&0&0&0&0&0&0&1&1&0&0&0&0&0&0 \\ 1&1&0&0&1&1&0&0&0&0&0&0&0&0&0&0 \\ 1&1&0&0&1&1&0&0&1&1&0&0&1&1&0&0 \\ 1&1&1&1&0&0&0&0&0&0&0&0&0&0&0&0 \\ 1&1&1&1&0&0&0&0&1&1&1&1&0&0&0&0 \\ 1&1&1&1&1&1&1&1&0&0&0&0&0&0&0&0 \\ 1&1&1&1&1&1&1&1&1&1&1&1&1&1&1&1 \end{bmatrix}.$$

Note: The inverse of the generator matrix equals to itself, i.e., $G = G^{-1}$, or $G \cdot G = I$.

Exercise: Construct the generator matrix G_{32} using the matrix G_{16}.

Example 1.7 If $F = \begin{bmatrix} 1 & 0 \\ 1 & 1 \end{bmatrix}$, find $F^{\otimes 2}$.

Solution 1.7 $F^{\otimes 2}$ is calculated as

$$F^{\otimes 2} = \begin{bmatrix} 1 \cdot F & 0 \cdot F \\ 1 \cdot F & 1 \cdot F \end{bmatrix} \rightarrow F^{\otimes 2} = \begin{bmatrix} 1&0&0&0 \\ 1&1&0&0 \\ 1&0&1&0 \\ 1&1&1&1 \end{bmatrix}.$$

Example 1.8 In some articles, we see that the generator matrix of the polar code is defined as

$$G_N = F^{\otimes n}$$

but it must have been defined as

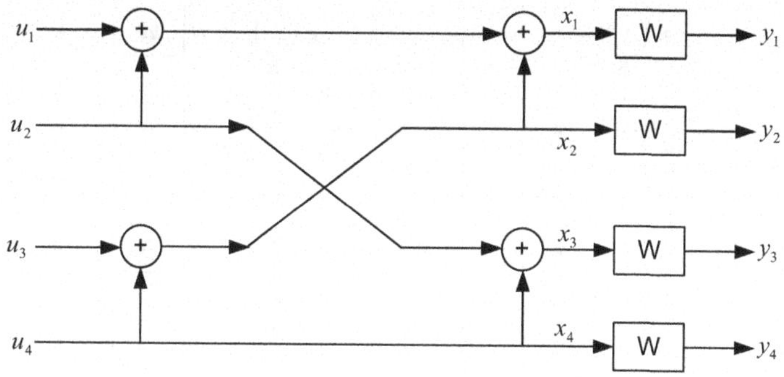

Fig. 1.18 Polar encoder for $N = 4$

$$G_N = B_N F^{\otimes n}.$$

Then, why do they define it as $G_N = F^{\otimes n}$?

Solution 1.8 Let's explain it by an example.

The generator matrix of the polar encoder shown in Fig. 1.18 for $N = 4$ is

$$G = \begin{bmatrix} 1\,0\,0\,0 \\ 1\,0\,1\,0 \\ 1\,1\,0\,0 \\ 1\,1\,1\,1 \end{bmatrix}$$

and the encoding is performed as

$$[x_1\, x_2\, x_3\, x_4] = [u_1\, u_2\, u_3\, u_4] \times \begin{bmatrix} 1\,0\,0\,0 \\ 1\,0\,1\,0 \\ 1\,1\,0\,0 \\ 1\,1\,1\,1 \end{bmatrix}.$$

The encoder structure in Fig. 1.18 can be redrawn as in Fig. 1.19. Considering Fig. 1.19, the encoding operation can be written as

$$[x_1\, x_2\, x_3\, x_4] = [u_1\, u_3\, u_2\, u_4] \times G'$$

where the generator matrix G' can be obtained from G by performing the permutations, which are done on the information sequence, i.e., $[u_1\, u_2\, u_3\, u_4] \rightarrow [u_1\, u_3\, u_2\, u_4]$, on columns of G, i.e., reorder the columns of G as $[c_1\, c_2\, c_3\, c_4] \rightarrow [c_1\, c_3\, c_2\, c_4]$ leading to

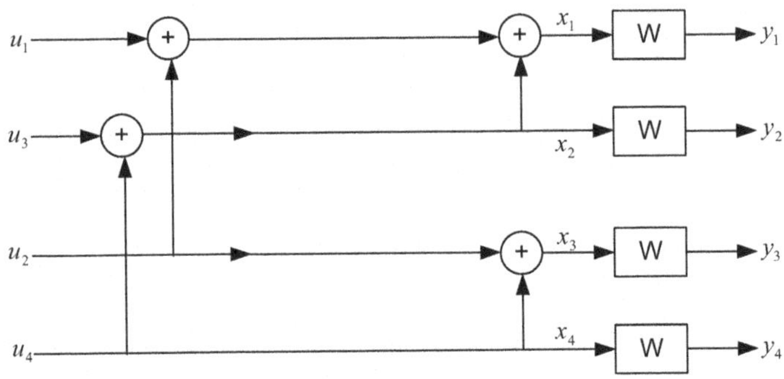

Fig. 1.19 Polar encoder for $N = 4$ with straight lines

$$G = \begin{bmatrix} 1\,0\,0\,0 \\ 1\,0\,1\,0 \\ 1\,1\,0\,0 \\ 1\,1\,1\,1 \end{bmatrix} \rightarrow G' = \begin{bmatrix} 1\,0\,0\,0 \\ 1\,1\,0\,0 \\ 1\,0\,1\,0 \\ 1\,1\,1\,1 \end{bmatrix}.$$

When G' is inspected, we see that it is equal to $F^{\otimes 2}$. Thus, we can calculate the generator matrix using only $F^{\otimes n}$, however, in this case the information vector to be encoded should be permuted, i.e., the encoding operation can be performed as

$$\bar{x} = \pi(\bar{u}) \cdot G' \rightarrow [x_1\, x_2 \ldots] = \pi([u_1\, u_2 \ldots]) \cdot F^{\otimes n}$$

On the other hand, some of the authors indicate the encoding operation as

$$\bar{x} = \bar{d} \cdot G' \rightarrow [x_1\, x_2 \ldots] = [d_1\, d_2 \ldots] \cdot F^{\otimes n}$$

which does not involve the permutation information in data vector. However, such an approach can be confusing for those who are new in the subject.

For this reason, we advise to the reader to always permutation information in the encoding operation, or just use the formula $G_N = B_N F^{\otimes n}$ for the generator matrix, since it automatically inserts the permutation information into the encoding operation.

Example 1.9 For $N = 8$, the generator matrices obtained using

$$G_8 = B_8 F^{\otimes 3} \quad G_8' = F^{\otimes 3}$$

are

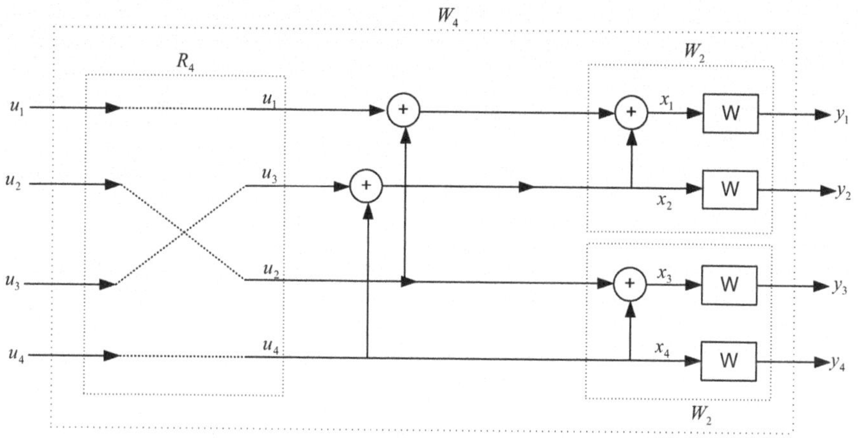

Fig. 1.20 Polar encoder for $N = 4$ using alternative general structure

$$G_8 = \begin{bmatrix} 1\,0\,0\,0\,0\,0\,0\,0 \\ 1\,0\,0\,0\,1\,0\,0\,0 \\ 1\,0\,1\,0\,0\,0\,0\,0 \\ 1\,0\,1\,0\,1\,0\,1\,0 \\ 1\,1\,0\,0\,0\,0\,0\,0 \\ 1\,1\,0\,0\,1\,1\,0\,0 \\ 1\,1\,1\,1\,0\,0\,0\,0 \\ 1\,1\,1\,1\,1\,1\,1\,1 \end{bmatrix} \quad G'_8 = \begin{bmatrix} 1\,0\,0\,0\,0\,0\,0\,0 \\ 1\,1\,0\,0\,0\,0\,0\,0 \\ 1\,0\,1\,0\,0\,0\,0\,0 \\ 1\,1\,1\,1\,0\,0\,0\,0 \\ 1\,0\,0\,0\,1\,0\,0\,0 \\ 1\,1\,0\,0\,1\,1\,0\,0 \\ 1\,0\,1\,0\,1\,0\,1\,0 \\ 1\,1\,1\,1\,1\,1\,1\,1 \end{bmatrix}. \tag{1.32}$$

If we write the generator matrix G_8 using its columns vectors as $G_8 = [\bar{c}_1\,\bar{c}_2\,\bar{c}_3\,\bar{c}_4\,\bar{c}_5\,\bar{c}_6\,\bar{c}_7\,\bar{c}_8]$, then from (1.32), it is clear that generator matrix G'_8 can be written as $G'_8 = [\bar{c}_1\,\bar{c}_5\,\bar{c}_3\,\bar{c}_7\,\bar{c}_2\,\bar{c}_6\,\bar{c}_4\,\bar{c}_8]$.

This means that if we do the encoding operation using the generator matrix G_8 as in

$$x_1^8 = [u_1\,u_2\,u_3\,u_4\,u_5\,u_6\,u_7\,u_8] \times G_8$$

then we can do the encoding operation using G'_8 as in

$$x_1^8 = [u_1\,u_5\,u_3\,u_7\,u_2\,u_6\,u_4\,u_8] \times G'_8$$

where the information sequence is permuted according to the permutation between columns of G_8 and G'_8.

Example 1.10 The polar encoder for $N = 4$ is depicted in Fig. 1.20. Is it possible to design another polar encoder structure equivalent to the one in Fig. 1.20?

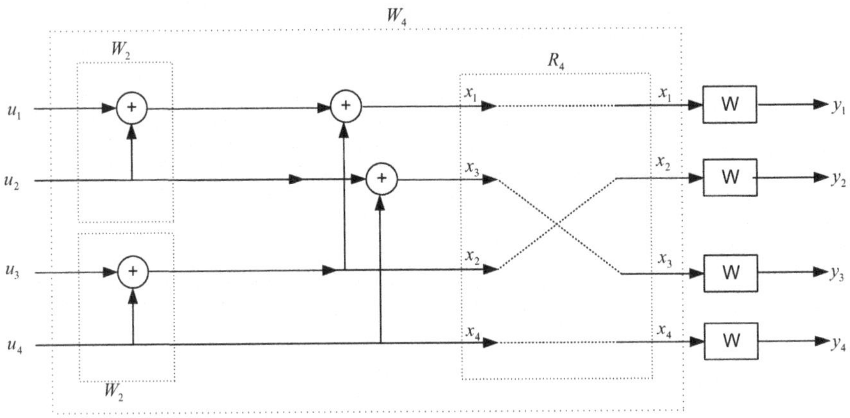

Fig. 1.21 Polar encoder for Example 1.10

Solution 1.10 From Fig. 1.20, we can write the relation between data-word u_1^4 and the code-word x_1^4 as in

$$x_1^4 = u_1^4 \times G_4. \tag{1.33}$$

The generator matrix in (1.33) has the property $G_4 = G_4^{-1}$, i.e., its inverse equals to itself. From (1.33), we can write

$$u_1^4 = x_1^4 \times G_4^{-1}$$

which implies that if we change the directions of the signal flows in (1.33), and change the places of u_1^4 and x_1^4, we can get an alternative polar encoder structure. The alternative polar encoder is depicted in Fig. 1.21.

Problems

(1) Draw the polar encoder structure for information vector length $N = 2$.
(2) Draw the polar encoder structure for information vector length $N = 4$.
(3) Draw the polar encoder structure for information vector length $N = 8$.
(4) Find the permutation matrix R_N used in polar encoder structures for $N = 2$, $N = 4$, $N = 8$, $N = 16$, and $N = 32$.
(5) Construct the polar encoder structure for $N = 4$ in a recursive manner, i.e. construct W_4 in a recursive manner using two W_2 and one R_4 using the **Method-1** explained in the book.
(6) Construct the polar encoder structure for $N = 4$ in a recursive manner, i.e. construct W_4 in a recursive manner using two W_2 and one R_4 using the **Method-2** explained in the book.

(7) Construct the polar encoder structure for $N = 8$ in a recursive manner, i.e. construct W_8 in a recursive manner using two W_4 and one R_8 using the **Method-1** explained in the book.
(8) Construct the polar encoder structure for $N = 8$ in a recursive manner, i.e. construct W_8 in a recursive manner using two W_4 and one R_8 using the **Method-2** explained in the book.
(9) Construct the polar encoder structure for $N = 16$ in a recursive manner, i.e. construct W_{16} in a recursive manner using two W_8 and one R_{16} using the **Method-1** explained in the book.
(10) Construct the polar encoder structure for $N = 16$ in a recursive manner, i.e. construct W_{16} in a recursive manner using two W_8 and one R_{16} using the **Method-2** explained in the book.
(11) Construct the polar encoder structure for $N = 32$ in a recursive manner, i.e. construct W_{32} in a recursive manner using two W_{16} and one R_{32} using the **Method-1** explained in the book.
(12) Construct the polar encoder structure for $N = 32$ in a recursive manner, i.e. construct W_{32} in a recursive manner using two W_{16} and one R_{32} using the **Method-2** explained in the book.
(13) If $F = \begin{bmatrix} 1 & 0 \\ 1 & 1 \end{bmatrix}$, find $F^{\otimes 2}$, $F^{\otimes 3}$, $F^{\otimes 4}$.
(14) Calculate the generator matrix G for $N = 4$, $N = 8$, $N = 16$, $N = 32$ mathematically.
(15) If the generator matrix G is available for $N = 32$, i.e., G_{32} is available. How do you construct G_{64} using G_{32}.
(16) Calculate the minimum distance of polar codes considering the generator matrices G_4, G_8, G_{16}, G_{32}.
(17) Consider the generator matrix G_{32}, write a matlab program to display the spectrum information for this polar code.
(18) What is the inverse of G_8?
(19) Find the parity check matrix of polar code for $N = 2$, $N = 4$, $N = 8$, $N = 16$, and $N = 32$.
(20) Find the systematic form of the generator matrix of polar codes for $N = 2$, $N = 4$, $N = 8$, $N = 16$, and $N = 32$.
(21) Construct the syndrome table of the polar codes for $N = 2$, $N = 4$, $N = 8$, $N = 16$, and $N = 32$.

Chapter 2
Decoding of Polar Codes

As we explained in Chap. 1, polar encoder structures can be constructed in a recursive manner. This leads to recursive decoding structures. In [1], Arıkan proposed successive cancellation decoding algorithm for the decoding of polar codes and derived recursive expression for the decoding operations.

Successive cancellation decoding can also be considered as bit-by-bit sequential decoding. In this decoding algorithm, bits are decoded in a sequential manner, and previously decoded bits are used for the decoding of current bit. That is why it is called successive cancelation decoding. In this chapter we will first explain the successive cancellation decoding logic and provide the necessary formulas, then explain the successive cancellation decoding operation with a tree structure proposed in [2]. It is easier to follow the decoding steps of the successive cancelation decoding algorithm with the tree structure [2].

2.1 Kernel Encoder and Decoder Units of the Polar Codes

In this section, we will inspect the kernel encoding and decoding units of the polar codes and derive the fundamental formulas necessary for the decoding operation.

The kernel unit, which can also be considered as the polar encoder unit with the smallest length code-words at its output, is depicted in Fig. 2.1 and it is repeatedly used in polar encoder structures. From Fig. 2.1, we can write that

In Fig. 2.1, $[a\ b]$ is the data-word encoded, and $[c\ d]$ is the code-word obtained from encoding of $[a\ b]$. The code bits are transmitted through the identical channels indicated by W. For the moment, let's assume that the bits c and d are transmitted through a discrete memoryless channel, and $[y_1\ y_2]$ are the signals at the receiver side At the decoder side, the flow of the signals change direction as shown in Fig. 2.2 where $\hat{c}$ and $\hat{d}$ are the estimated code bits. Recovering data bits from the code-bits at the receiver side is called decoding operation. In Fig. 2.2, we aim to decide on the

O. Gazi, *Polar Codes*, Springer Topics in Signal Processing 15,
https://doi.org/10.1007/978-981-13-0737-9_2

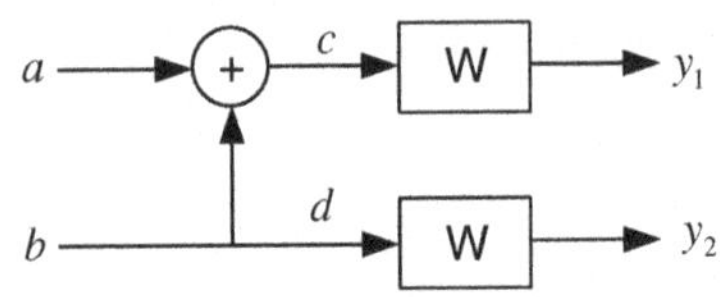

Fig. 2.1 Kernel encoder unit with two DMCs

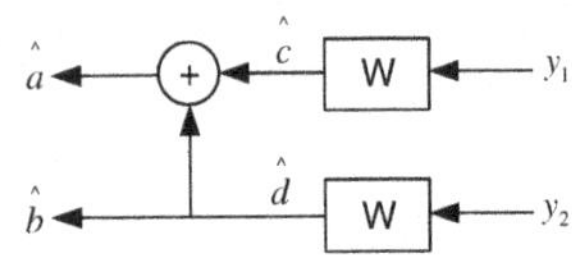

Fig. 2.2 Kernel decoder unit

data bits $[\hat{a}\,\hat{b}]$, using the code bits $[\hat{c}\,\hat{d}]$. In the sequel, we will explain the decoding operation for the kernel unit in Fig. 2.2.

The communication channel W is nothing but a probability matrix between channel inputs and outputs. Estimating code bits is nothing but determining the channel input probabilities, i.e., determining the probabilities $P(\hat{c}=0)$, $P(\hat{c}=1)$, $P(\hat{d}=0)$, $P(\hat{d}=1)$ whose calculations depends on the channel transition probabilities and channel outputs. Decoding of the data bits are done in a sequential manner, i.e., first the value of $\hat{a}$ is evaluated, then using the value of $\hat{a}$ and the received signals y_1 and y_2 the value of $\hat{b}$ is evaluated.

For the decoding unit in Fig. 2.2, we can write

$$\hat{a} = \hat{c} \oplus \hat{d} \quad \hat{b} = \hat{d}. \tag{2.1}$$

It is clear from (2.1) that $\hat{a}=0$, if $\hat{c}=0$ and $\hat{d}=0$ or $\hat{c}=1$ and $\hat{d}=1$. This means that

$$P(\hat{a}=0) = P(\hat{c}=0)P\left(\hat{d}=0\right) + P(\hat{c}=1)P\left(\hat{d}=1\right) \tag{2.2}$$

In a similar manner, $\hat{a}=1$ if $\hat{c}=0$ and $\hat{d}=1$ or $\hat{c}=1$ and $\hat{d}=0$. This leads to

$$P(\hat{a}=1) = P(\hat{c}=0)P\left(\hat{d}=1\right) + P(\hat{c}=1)P\left(\hat{d}=0\right). \tag{2.3}$$

Let's define the $LR(\hat{a})$, i.e., likelihood ratio of $\hat{a}$, as

$$LR(\hat{a}) = \frac{P(\hat{a}=0)}{P(\hat{a}=1)} \tag{2.4}$$

in which using (2.2) and (2.3), we get

$$LR(\hat{a}) = \frac{P(\hat{c}=0)P\left(\hat{d}=0\right) + P(\hat{c}=1)P\left(\hat{d}=1\right)}{P(\hat{c}=0)P\left(\hat{d}=1\right) + P(\hat{c}=1)P\left(\hat{d}=0\right)} \tag{2.5}$$

Dividing the numerator and denominator of (2.5) by $P(\hat{c}=1)P(\hat{d}=1)$, we obtain

$$LR(\hat{a}) = \frac{1+\frac{P(\hat{c}=0)P(\hat{d}=0)}{P(\hat{c}=1)P(\hat{d}=1)}}{\frac{P(\hat{d}=0)}{P(\hat{d}=1)}+\frac{P(\hat{c}=0)}{P(\hat{c}=1)}} \tag{2.6}$$

which can be written as

$$LR(\hat{a}) = \frac{1+LR(\hat{c})LR(\hat{d})}{LR(\hat{c})+LR(\hat{d})}. \tag{2.7}$$

Assume that first the value of $\hat{a}$ is determined. Then, we want to determine the value of $\hat{b}$. We can determine the value of $\hat{b}$ as follows.

From Fig. 2.2, it is clear that, if $\hat{a}=0$, then $\hat{b}=0$ with the constraint $\hat{c}=0$ and $\hat{d}=0$, and $\hat{b}=1$ with the constraint $\hat{c}=1$ and $\hat{d}=1$. This means that

$$P_{\hat{a}=0}(\hat{b}=0) = P(\hat{c}=0)P(\hat{d}=0) \quad P_{\hat{a}=0}(\hat{b}=1) = P(\hat{c}=1)P(\hat{d}=1). \tag{2.8}$$

We define the $LR_{\hat{a}=0}(\hat{b})$ as

$$LR_{\hat{a}=0}(\hat{b}) = \frac{P_{\hat{a}=0}(\hat{b}=0)}{P_{\hat{a}=0}(\hat{b}=1)} \tag{2.9}$$

in which using (2.8), we obtain

$$LR_{\hat{a}=0}(\hat{b}) = \frac{P(\hat{c}=0)P(\hat{d}=0)}{P(\hat{c}=1)P(\hat{d}=1)} \tag{2.10}$$

which can be written as

$$LR_{\hat{a}=0}(\hat{b}) = LR(\hat{c})LR(\hat{d}). \tag{2.11}$$

In a similar manner, for the case of $\hat{a}=1$, we can determine the value of $\hat{b}$ as follows. If $\hat{a}=1$, then $\hat{b}=0$ with the constraint $\hat{c}=1$ and $\hat{d}=0$, and $\hat{b}=1$ with the constraint $\hat{c}=0$ and $\hat{d}=1$. This means that

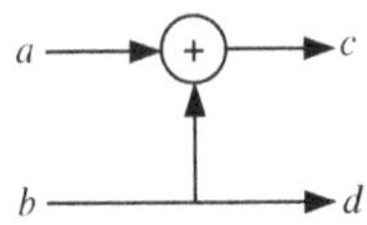

Fig. 2.3 Kernel encoder unit for Example 2.1

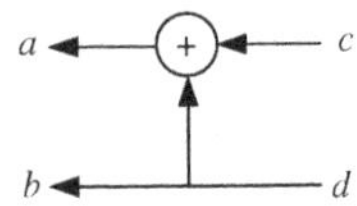

Fig. 2.4 Kernel decoder unit for Example 2.1

$$P_{\hat{a}=1}\left(\hat{b}=0\right)=P(\hat{c}=1)P\left(\hat{d}=0\right)\quad P_{\hat{a}=1}\left(\hat{b}=1\right)=P(\hat{c}=0)P\left(\hat{d}=1\right). \tag{2.12}$$

We define the $LR_{\hat{a}=1}(\hat{b})$ as

$$LR_{\hat{a}=1}\left(\hat{b}\right)=\frac{P_{\hat{a}=1}\left(\hat{b}=0\right)}{P_{\hat{a}=1}\left(\hat{b}=1\right)} \tag{2.13}$$

in which using (2.12), we obtain

$$LR_{\hat{a}=1}\left(\hat{b}\right)=\frac{P(\hat{c}=1)P\left(\hat{d}=0\right)}{P(\hat{c}=0)P\left(\hat{d}=1\right)} \tag{2.14}$$

which can be written as

$$LR_{\hat{a}=1}\left(\hat{b}\right)=\left[LR(\hat{c})\right]^{-1}LR\left(\hat{d}\right). \tag{2.15}$$

We can combine (2.11) and (2.15) under a single term as

$$LR\left(\hat{b}\right)=\left[LR(\hat{c})\right]^{1-2\hat{a}}LR\left(\hat{d}\right). \tag{2.16}$$

Note: For easiness of the notation, we will not use hat for the symbols in the decoder structure, i.e., we will use a for $\hat{a}$, b for $\hat{b}$, c for $\hat{c}$ and so on, unless otherwise is indicated.

Example 2.1 Consider the encoding unit in Fig. 2.3 where a and b data bits, c and d are code bits. Assume that c and d are transmitted through a discrete memoryless channel.

At the receiver side we have the decoder block shown in Fig. 2.4 where

$$LR(c)=\frac{P(c=0)}{P(c=1)}=10\quad LR(d)=\frac{P(d=0)}{P(d=1)}=0.1.$$

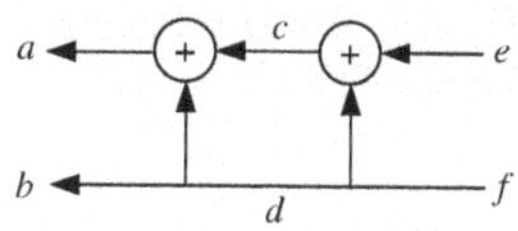

Fig. 2.5 Decoder unit for Exercise 2.1

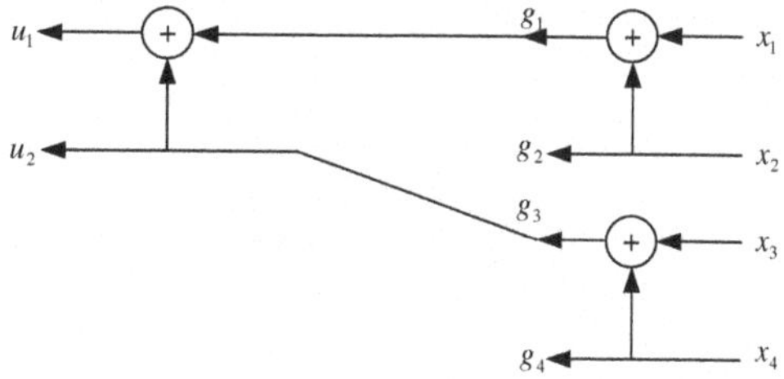

Fig. 2.6 Decoder segment for Exercise 2.2

First calculate $LR(a)$, and decode a, then calculate $LR(b)$ and decide on the value of b.

Solution 2.1 $LR(a)$ can be calculated as

$$LR(a) = \frac{1 + LR(c)LR(d)}{LR(c) + LR(d)} \rightarrow LR(a) = \frac{1 + 10 \times 0.1}{10 + 0.1} \rightarrow LR(a) \approx 0.2.$$

From $LR(a) \approx 0.2$, we can decide the value of bit a as 1. After determination of first data bit, we can start to the decoding of second data bit b whose $LR(b)$ can be calculated as

$$LR(b) = [LR(c)]^{1-2a} LR(d). \tag{2.17}$$

Since $a = 1$, Eq. (2.17) is evaluated as

$$LR(b) = [LR(c)]^{1-2a} LR(d) \rightarrow LR(b) = 10^{1-2\times 1} \times 0.1 \rightarrow LR(b) = 0.01.$$

From which we can decide on the value of b as 1. Thus, we found the data word ab as 11.

Exercise 2.1 *For the decoder structure in Fig. 2.5, if* $LR(e) = 10$ *and* $LR(f) = 0.1$*, calculate* $LR(c)$, $LR(d)$, $LR(a)$*, and* $LR(b)$*.*

Exercise 2.2 *For the decoder structure in Fig. 2.6, if* $LR(x_1) = 10, LR(x_2) = 10, LR(x_3) = 10$ *and* $LR(x_4) = 0.1$*, calculate* $LR(g_1)$*,* $LR(g_2)$*,* $LR(g_3), LR(g_4)$*,* $LR(u_1)$*, and* $LR(u_2)$*. Using* $LR(u_1)$*, and* $LR(u_2)$ *determine the bit values of* u_1 *and* u_2*.*

Exercise 2.3 *For the decoder structure in Fig. 2.7, if* $LR(x_1) = 1, LR(x_2) = 10, LR(x_3) = 0.1$ *and* $LR(x_4) = 10$*, calculate* $LR(g_1), LR(g_2), LR(g_3), LR(g_4)$*,* $LR(u_3)$*, and* $LR(u_4)$*. Using* $LR(u_3)$*, and* $LR(u_4)$ *determine the bit values of* u_3 *and* u_4*.*

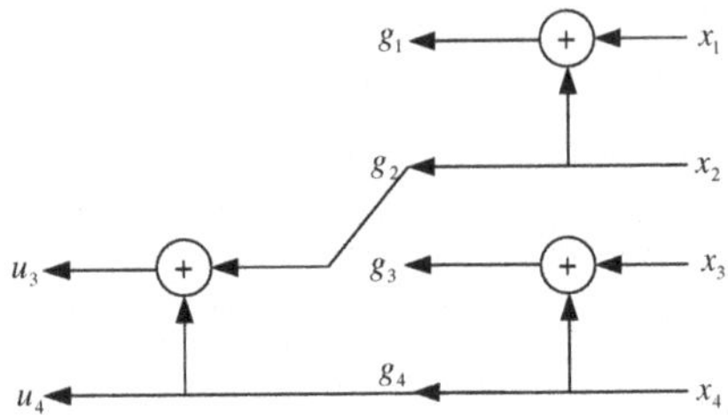

Fig. 2.7 Decoder segment for Exercise 2.3

2.2 Decoding Tree for the Successive Cancelation Decoding of Polar Codes

In this section, we will explain the successive cancelation decoding of polar codes using a decoding tree [2]. For the simplicity of the illustration, let's consider the polar encoder in Fig. 2.8.

Once the encoded bits are transmitted, the decoding operation follows the reverse paths of Fig. 2.8 as shown in Fig. 2.9.

Consider the decoding of the first information bit, i.e., u_1. The decoding path of u_1 is depicted with bold line in Fig. 2.10.

The decoding path shown by bold lines in Fig. 2.10 can be redrawn in a tree structure as in Fig. 2.11.

In Fig. 2.10 or 2.11, the likelihood ratios are calculated as

$$LR(g_1) = \frac{1 + LR(x_1)LR(x_2)}{LR(x_1) + LR(x_2)} \quad LR(g_2) = \frac{1 + LR(x_3)LR(x_4)}{LR(x_3) + LR(x_4)}$$
$$LR(u_1) = \frac{1 + LR(g_1)LR(g_2)}{LR(g_1) + LR(g_2)} \tag{2.18}$$

where $LR(x_i), i = 1, \ldots, 4$ are evaluated considering the channel outputs $y_i, i = 1, \ldots, 4$ and channel transition probabilities. After deciding on the value of u_1, we

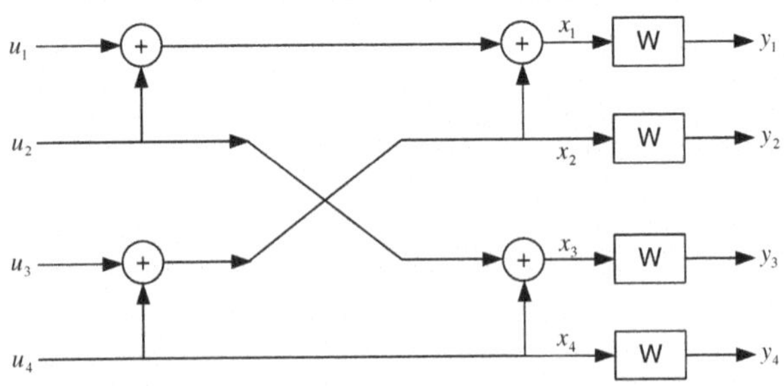

Fig. 2.8 Polar encoder for $N = 4$ with DMCs

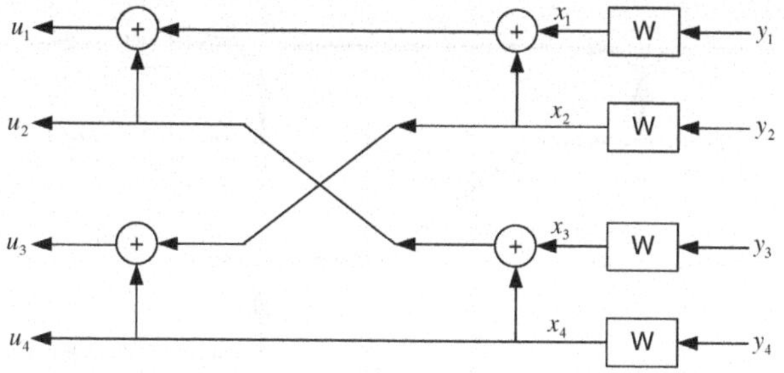

Fig. 2.9 Polar decoder for $N = 4$

Fig. 2.10 Decoding path for u_1

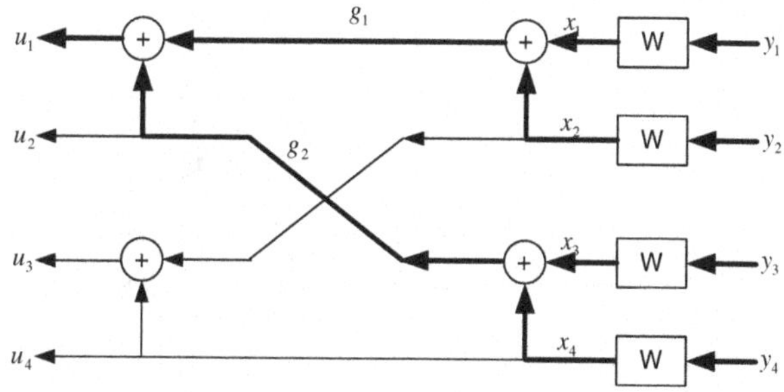

Fig. 2.11 Decoding tree for u_1

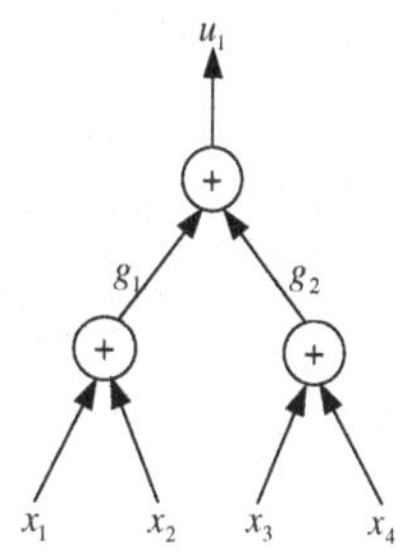

decode the information bit u_2. The decoding path for u_2 is depicted in Fig. 2.12 with bold lines.

The decoding path of u_2 shown by bold lines in Fig. 2.12 can be redrawn in a tree structure as in Fig. 2.13.

In Fig. 2.13, the likelihood ratios for g_1 and g_2 are calculated as

$$LR(g_1) = \frac{1 + LR(x_1)LR(x_2)}{LR(x_1) + LR(x_2)} \quad LR(g_2) = \frac{1 + LR(x_3)LR(x_4)}{LR(x_3) + LR(x_4)} \tag{2.19}$$

Since u_1 is known, $LR(u_2)$ can be calculated as

$$LR(u_2) = [LR(g_1)]^{1-2u_1} \times LR(g_2). \tag{2.20}$$

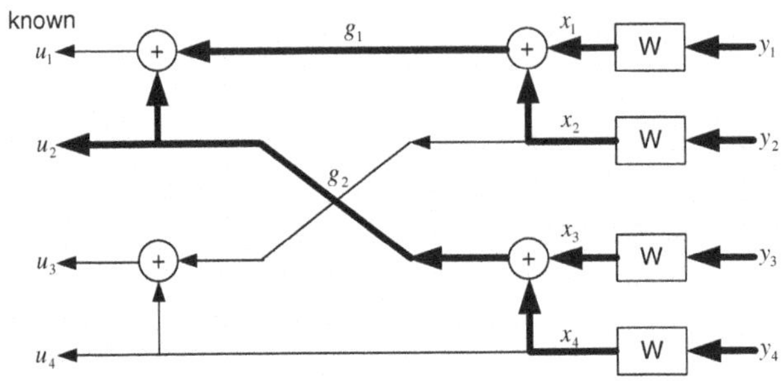

Fig. 2.12 Decoding path for u_2

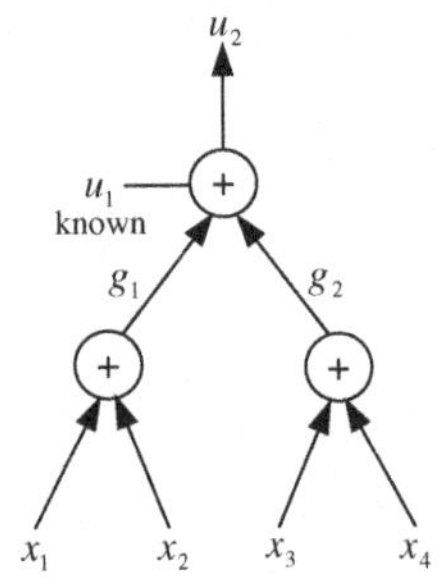

Fig. 2.13 Decoding tree for u_2

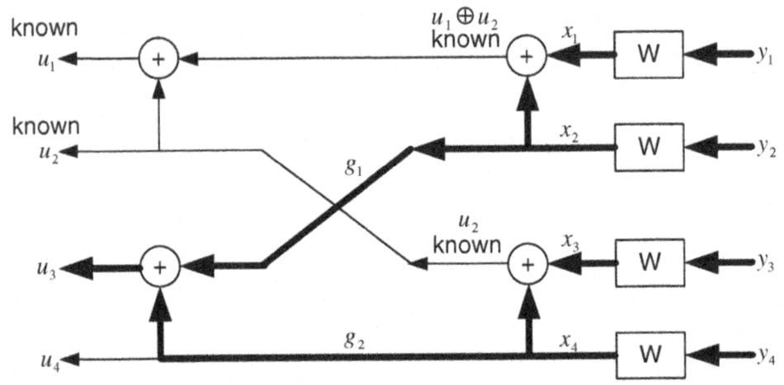

Fig. 2.14 Decoding path for u_3

After deciding u_2, we can decode data bit u_3. The decoding path for u_3 is depicted with bold lines in Fig. 2.14.

The decoding path of u_3 shown by bold lines in Fig. 2.14 can be redrawn in a tree structure as in Fig. 2.15.

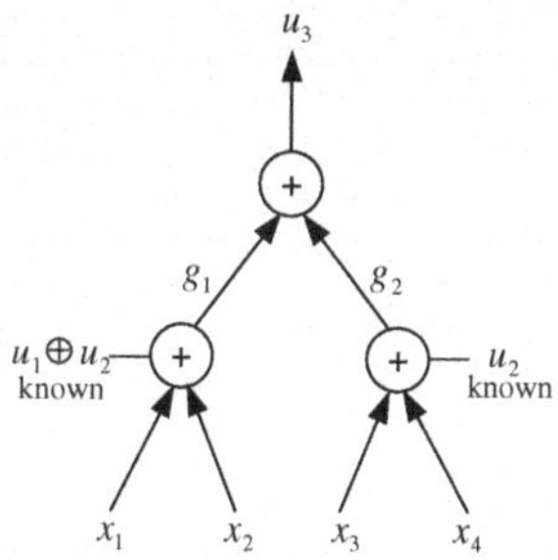

Fig. 2.15 Decoding tree for u_3

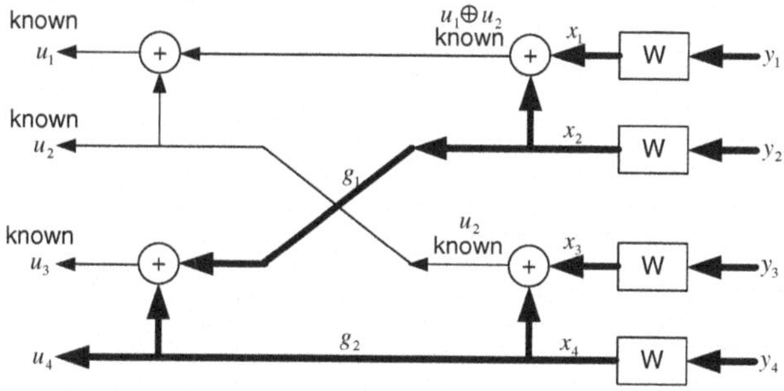

Fig. 2.16 Decoding path for u_4

In Fig. 2.15, the likelihood ratios for g_1, g_2 and u_3 are calculated as

$$LR(g_1) = [LR(x_1)]^{1-2(u_1 \oplus u_2)} \times LR(x_2) \quad LR(g_2) = [LR(x_3)]^{1-2u_2} \times LR(x_4)$$
$$LR(u_3) = \frac{1 + LR(g_1)LR(g_2)}{LR(g_1) + LR(g_2)}. \tag{2.21}$$

After decoding the information bit u_3, let's consider the decoding of u_4. The decoding path for u_4 is depicted with bold lines in Fig. 2.16.

The decoding path of u_3 shown by bold lines in Fig. 2.16 can be redrawn in a tree structure as in Fig. 2.17.

In Fig. 2.17, the likelihood ratios for g_1, g_2 and u_4 are calculated as

$$LR(g_1) = [LR(x_1)]^{1-2(u_1 \oplus u_2)} \times LR(x_2) \quad LR(g_2) = [LR(x_3)]^{1-2u_2} \times LR(x_4)$$
$$LR(u_4) = [LR(g_1)]^{1-2u_3} \times LR(g_2). \tag{2.22}$$

For better understanding of the decoding operation of the polar codes, let's consider the case for $N = 8$.

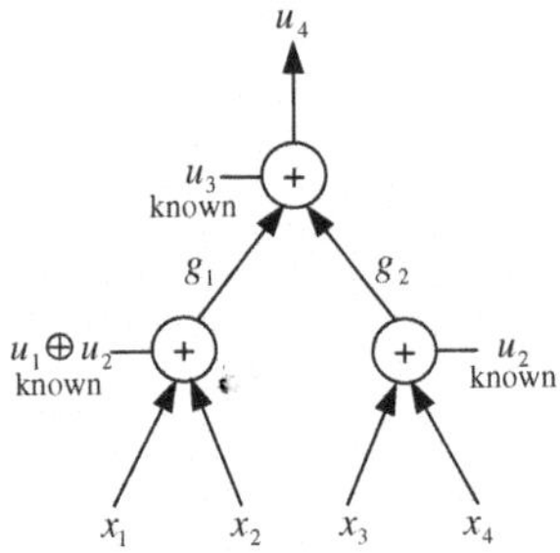

Fig. 2.17 Decoding tree for u_4

Fig. 2.18 Polar decoder for $N = 8$

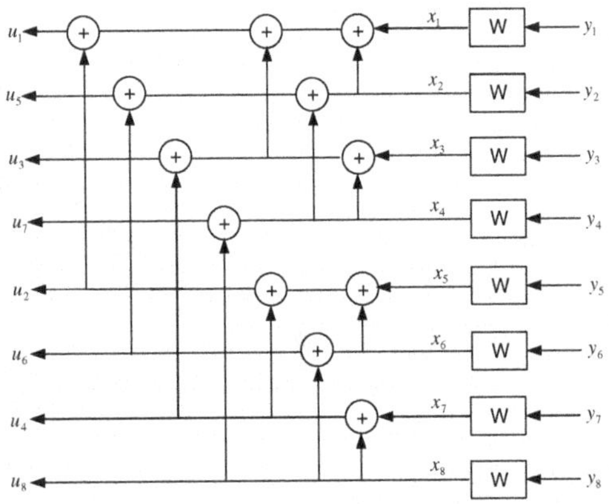

The signal flow for the decoding of information bits is depicted in Fig. 2.18. The decoding of the information bit u_1 is depicted in Fig. 2.19 with bold lines.

The decoding path shown in Fig. 2.19 can be expressed using a tree as in Fig. 2.20. In Fig. 2.20, the likelihood ratios for $g_i, i = 1, \ldots, 4$ can be calculated as

$$LR(g_1) = \frac{1 + LR(x_1)LR(x_2)}{LR(x_1) + LR(x_2)} \quad LR(g_2) = \frac{1 + LR(x_3)LR(x_4)}{LR(x_3) + LR(x_4)}$$
$$LR(g_3) = \frac{1 + LR(x_5)LR(x_6)}{LR(x_5) + LR(x_6)} \quad LR(g_4) = \frac{1 + LR(x_7)LR(x_8)}{LR(x_7) + LR(x_8)}. \tag{2.23}$$

The likelihood ratios for h_1 and h_2 can be calculated as

$$LR(h_1) = \frac{1 + LR(g_1)LR(g_2)}{LR(g_1) + LR(g_2)} \quad LR(h_2) = \frac{1 + LR(g_3)LR(g_4)}{LR(g_3) + LR(g_4)}. \tag{2.24}$$

The likelihood ratio for information bit u_1 is calculated as

$$LR(u_1) = \frac{1 + LR(h_1)LR(h_2)}{LR(h_1) + LR(h_2)}. \tag{2.25}$$

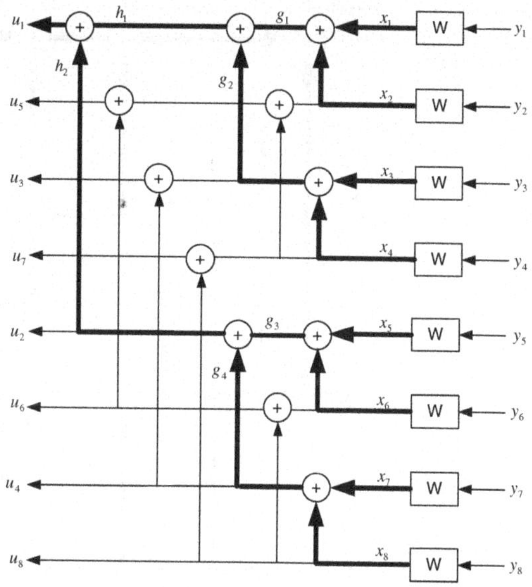

Fig. 2.19 Decoding path of u_1 when $N = 8$

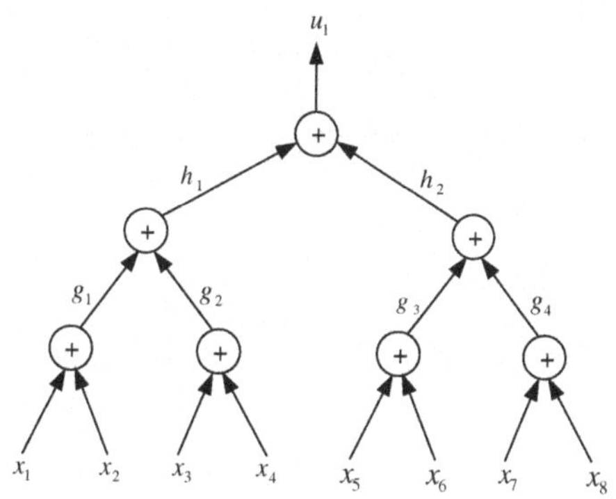

Fig. 2.20 Decoding tree of u_1 when $N = 8$

Once u_1 is decoded, we proceed with the decoding of u_2. The decoding path for u_2 is shown with the bold lines in Fig. 2.21.

The decoding path shown in Fig. 2.21 can be expressed using a tree as in Fig. 2.22. In Fig. 2.22, the likelihood ratios for $g_i, i = 1, \cdots, 4$ can be calculated as

$$LR(g_1) = \frac{1 + LR(x_1)LR(x_2)}{LR(x_1) + LR(x_2)} \quad LR(g_2) = \frac{1 + LR(x_3)LR(x_4)}{LR(x_3) + LR(x_4)}$$
$$LR(g_3) = \frac{1 + LR(x_5)LR(x_6)}{LR(x_5) + LR(x_6)} \quad LR(g_4) = \frac{1 + LR(x_7)LR(x_8)}{LR(x_7) + LR(x_8)}. \tag{2.26}$$

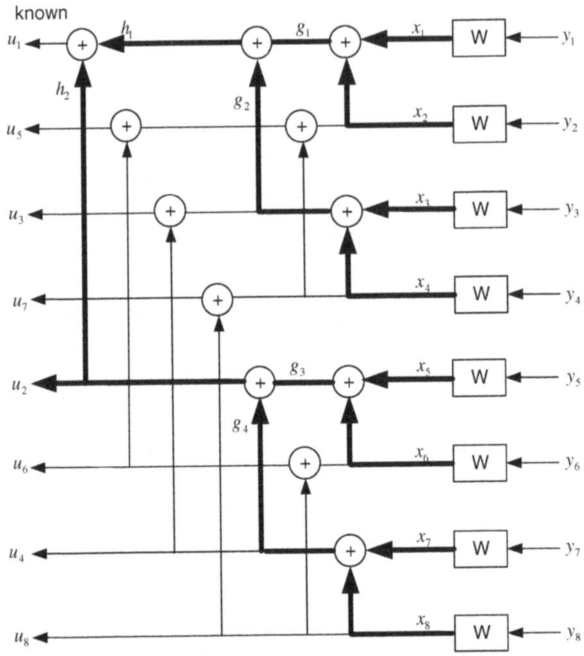

Fig. 2.21 Decoding path of u_2 when $N = 8$

Fig. 2.22 Decoding tree of u_2 when $N = 8$

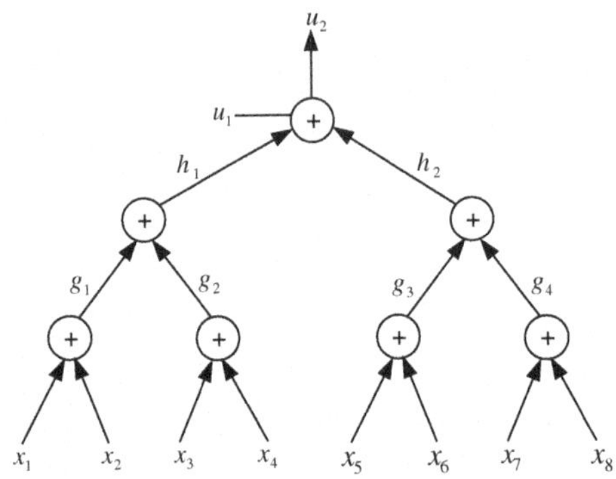

The likelihood ratios for h_1 and h_2 can be calculated as

$$LR(h_1) = \frac{1 + LR(g_1)LR(g_2)}{LR(g_1) + LR(g_2)} \quad LR(h_2) = \frac{1 + LR(g_3)LR(g_4)}{LR(g_3) + LR(g_4)}. \tag{2.27}$$

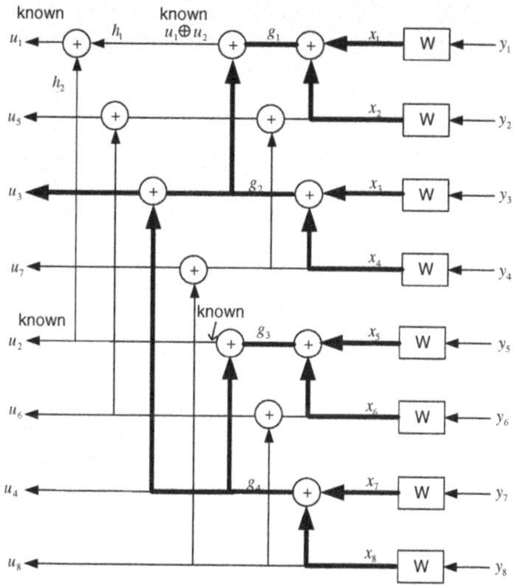

Fig. 2.23 Decoding path of u_3 when $N = 8$

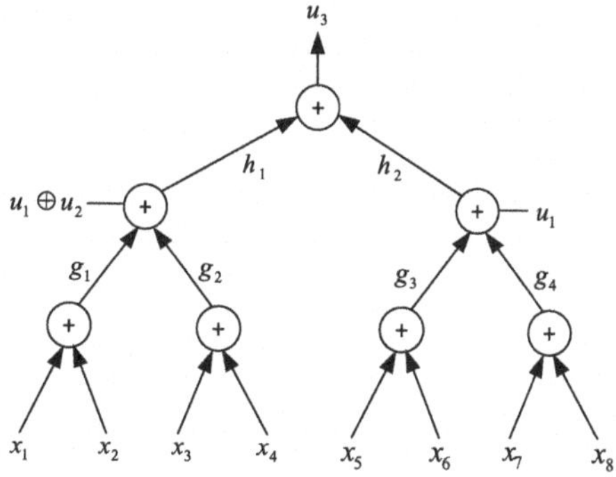

Fig. 2.24 Decoding tree of u_3 when $N = 8$

Using the previously solved bit u_1, the likelihood ratio for information bit u_2 is calculated as

$$LR(u_1) = [LR(h_1)]^{1-2u_1} LR(h_2). \tag{2.28}$$

After determining u_2, we can decode the information bit u_3 using the path shown in bold lines in Fig. 2.23.

The decoding path shown in Fig. 2.23 can be expressed using a tree as in Fig. 2.24. In Fig. 2.24, the likelihood ratios for g_i, $i = 1, \ldots, 4$ can be calculated as

$$\begin{aligned} LR(g_1) &= \frac{1 + LR(x_1)LR(x_2)}{LR(x_1) + LR(x_2)} & LR(g_2) &= \frac{1 + LR(x_3)LR(x_4)}{LR(x_3) + LR(x_4)} \\ LR(g_3) &= \frac{1 + LR(x_5)LR(x_6)}{LR(x_5) + LR(x_6)} & LR(g_4) &= \frac{1 + LR(x_7)LR(x_8)}{LR(x_7) + LR(x_8)}. \end{aligned} \tag{2.29}$$

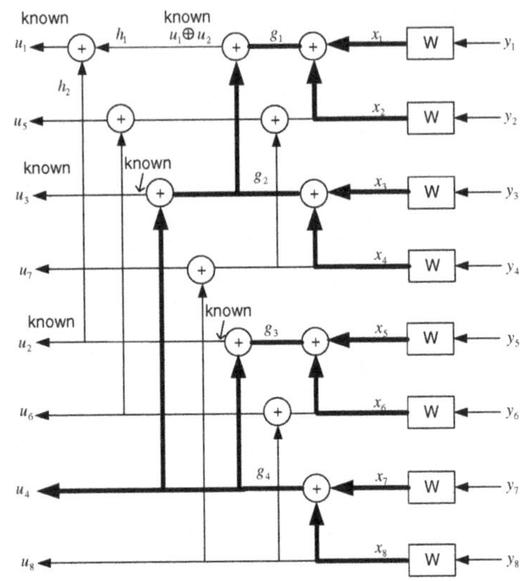

Fig. 2.25 Decoding path of u_4 when $N = 8$

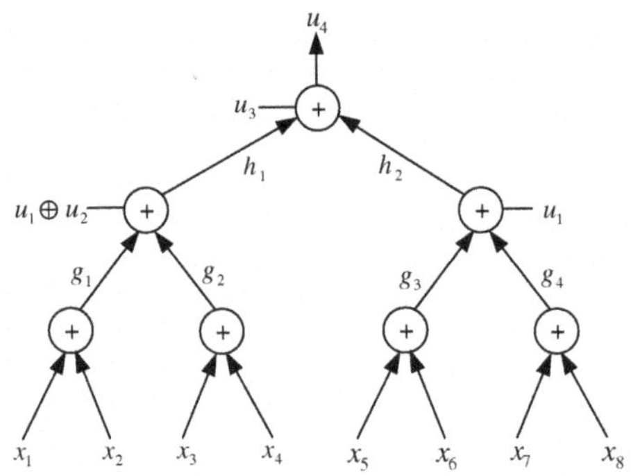

Fig. 2.26 Decoding tree of u_4 when $N = 8$

The likelihood ratios for h_1 and h_2 can be calculated as

$$LR(h_1) = [LR(g_1)]^{1-2(u_1 \oplus u_2)} LR(g_2) \quad LR(h_2) = [LR(g_3)]^{1-2u_1} LR(g_4). \quad (2.30)$$

The likelihood ratio for information bit u_3 is calculated as

$$LR(u_3) = \frac{1 + LR(h_1)LR(h_2)}{LR(h_1) + LR(h_2)}. \quad (2.31)$$

After deciding on the value of u_3, we proceed with the decoding of u_4. The decoding path of u_4 is shown with bold lines in Fig. 2.25.

The decoding path shown in Fig. 2.25 can be expressed using a tree as in Fig. 2.26.

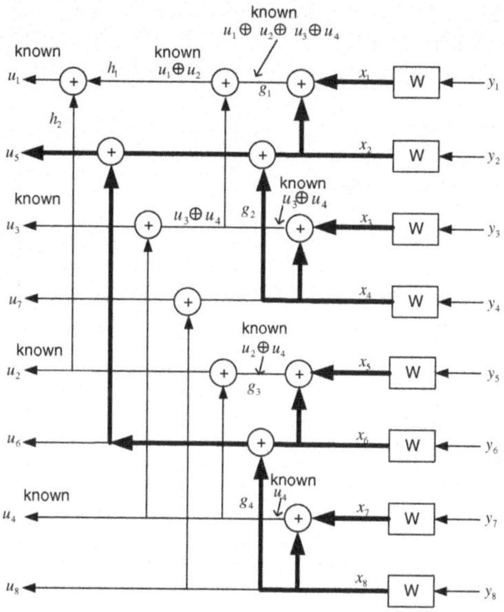

Fig. 2.27 Decoding path of u_5 when $N = 8$

In Fig. 2.26, the likelihood ratios for $g_i, i = 1, \ldots, 4$ can be calculated as

$$LR(g_1) = \frac{1 + LR(x_1)LR(x_2)}{LR(x_1) + LR(x_2)} \quad LR(g_2) = \frac{1 + LR(x_3)LR(x_4)}{LR(x_3) + LR(x_4)}$$
$$LR(g_3) = \frac{1 + LR(x_5)LR(x_6)}{LR(x_5) + LR(x_6)} \quad LR(g_4) = \frac{1 + LR(x_7)LR(x_8)}{LR(x_7) + LR(x_8)}. \tag{2.32}$$

The likelihood ratios for h_1 and h_2 can be calculated as

$$LR(h_1) = [LR(g_1)]^{1-2(u_1 \oplus u_2)} LR(g_2) \quad LR(h_2) = [LR(g_3)]^{1-2u_1} LR(g_4). \tag{2.33}$$

The likelihood ratio for information bit u_4 is calculated as

$$LR(u_4) = [LR(h_1)]^{1-2u_3} LR(h_2). \tag{2.34}$$

After deciding on the value of u_4, we proceed with the decoding of the information bit u_5. The decoding path of u_5 is shown with bold lines in Fig. 2.27.

The decoding path shown in Fig. 2.27 can be expressed using a tree as in Fig. 2.28.

In Fig. 2.28, the likelihood ratios for $g_i, i = 1, \ldots, 4$ can be calculated as

$$LR(g_1) = [LR(x_1)]^{1-2(u_1 \oplus u_2 \oplus u_3 \oplus u_4)} LR(x_2) \quad LR(g_2) = [LR(x_3)]^{1-2(u_3 \oplus u_4)} LR(x_4)$$
$$LR(g_3) = [LR(x_5)]^{1-2(u_2 \oplus u_4)} LR(x_6) \quad LR(g_4) = [LR(x_7)]^{1-2u_4} LR(x_8). \tag{2.35}$$

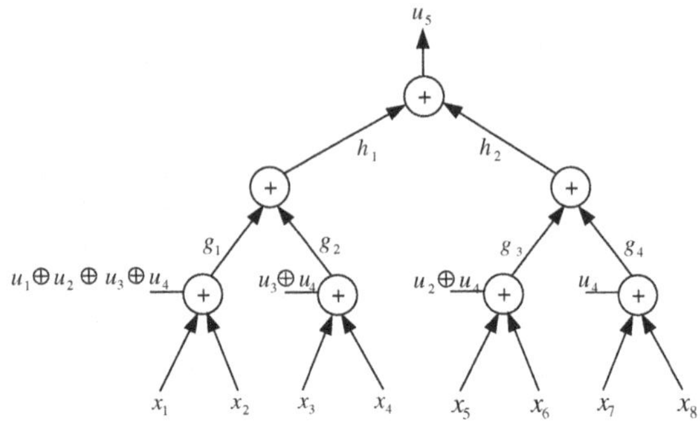

Fig. 2.28 Decoding tree of u_5 when $N = 8$

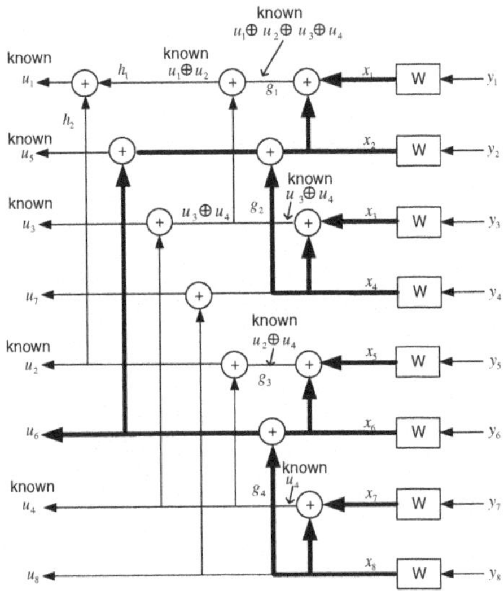

Fig. 2.29 Decoding path of u_6 when $N = 8$

The likelihood ratios for h_1 and h_2 can be calculated as

$$LR(h_1) = \frac{1 + LR(g_1)LR(g_2)}{LR(g_1) + LR(g_2)} \quad LR(h_2) = \frac{1 + LR(g_3)LR(g_4)}{LR(g_3) + LR(g_4)}. \tag{2.36}$$

The likelihood ratio for information bit u_5 is calculated as

$$LR(u_5) = \frac{1 + LR(h_1)LR(h_2)}{LR(h_1) + LR(h_2)}. \tag{2.37}$$

After deciding on the value of u_5, we proceed with the decoding of the information bit u_6. The decoding path of u_6 is shown with bold lines in Fig. 2.29.

The decoding path shown in Fig. 2.29 can be expressed using a tree as in Fig. 2.30.

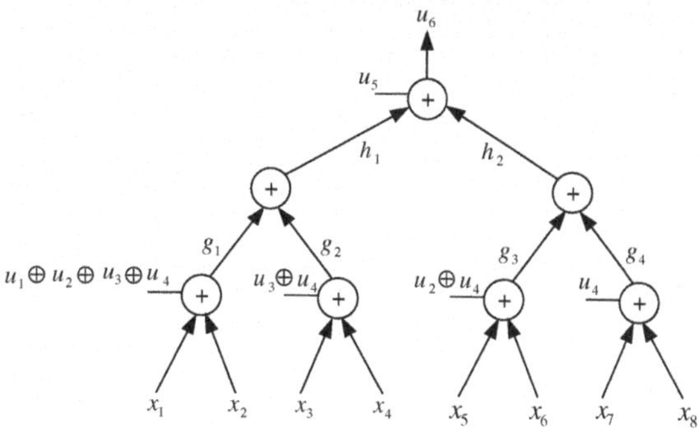

Fig. 2.30 Decoding path of u_6 when $N = 8$

In Fig. 2.30, the likelihood ratios for $g_i, i = 1, \ldots, 4$ can be calculated as

$$LR(g_1) = [LR(x_1)]^{1-2(u_1 \oplus u_2 \oplus u_3 \oplus u_4)} LR(x_2) \quad LR(g_2) = [LR(x_3)]^{1-2(u_3 \oplus u_4)} LR(x_4)$$
$$LR(g_3) = [LR(x_5)]^{1-2(u_2 \oplus u_4)} LR(x_6) \quad LR(g_4) = [LR(x_7)]^{1-2u_4} LR(x_8). \tag{2.38}$$

The likelihood ratios for h_1 and h_2 can be calculated as

$$LR(h_1) = \frac{1 + LR(g_1)LR(g_2)}{LR(g_1) + LR(g_2)} \quad LR(h_2) = \frac{1 + LR(g_3)LR(g_4)}{LR(g_3) + LR(g_4)}. \tag{2.39}$$

The likelihood ratio for information bit u_6 is calculated as

$$LR(u_6) = [LR(h_1)]^{1-2u_5} LR(h_2). \tag{2.40}$$

After deciding on the value of u_6, we proceed with the decoding of the information bit u_7. The decoding path of u_7 is shown with bold lines in Fig. 2.31.

The decoding path shown in Fig. 2.31 can be expressed using a tree as in Fig. 2.32. In Fig. 2.32, the likelihood ratios for $g_i, i = 1, \ldots, 4$ can be calculated as

$$LR(g_1) = [LR(x_1)]^{1-2(u_1 \oplus u_2 \oplus u_3 \oplus u_4)} LR(x_2) \quad LR(g_2) = [LR(x_3)]^{1-2(u_3 \oplus u_4)} LR(x_4)$$
$$LR(g_3) = [LR(x_5)]^{1-2(u_2 \oplus u_4)} LR(x_6) \quad LR(g_4) = [LR(x_7)]^{1-2u_4} LR(x_8). \tag{2.41}$$

The likelihood ratios for h_1 and h_2 can be calculated as

$$LR(h_1) = [LR(g_1)]^{1-2(u_5 \oplus u_6)} LR(g_2) \quad LR(h_2) = [LR(g_3)]^{1-2u_6} LR(g_4). \tag{2.42}$$

The likelihood ratio for information bit u_7 is calculated as

$$LR(u_7) = \frac{1 + LR(h_1)LR(h_2)}{LR(h_1) + LR(h_2)}. \tag{2.43}$$

After deciding on the value of u_7, we proceed with the decoding of the information bit u_8. The decoding path of u_8 is shown with bold lines in Fig. 2.33.

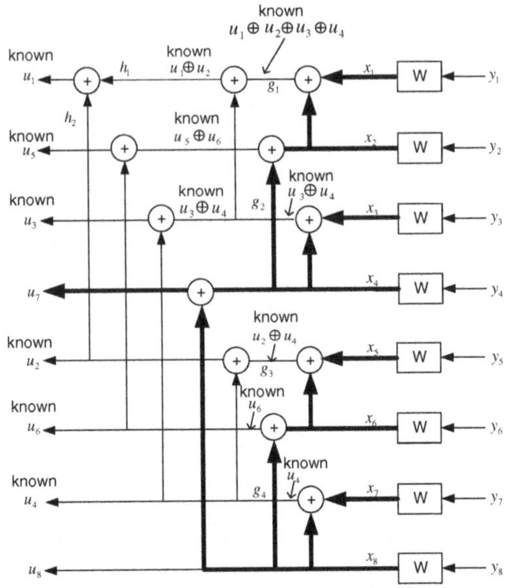

Fig. 2.31 Decoding path of u_7 when $N = 8$

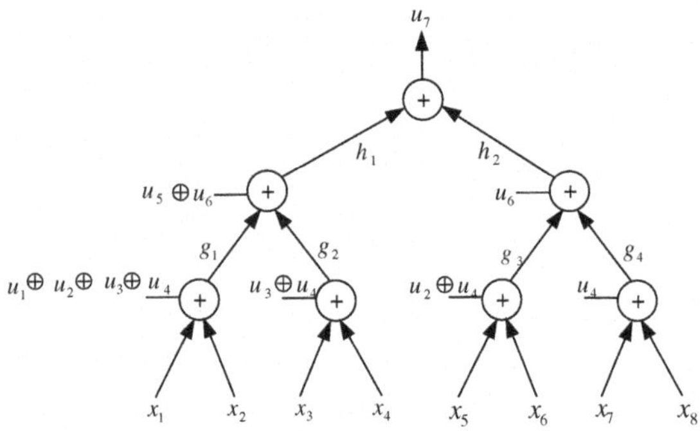

Fig. 2.32 Decoding tree for u_7 when $N = 8$

The decoding path shown in Fig. 2.33 can be expressed using a tree as in Fig. 2.34. In Fig. 2.34, the likelihood ratios for $g_i, i = 1, \ldots, 4$ can be calculated as

$$LR(g_1) = [LR(x_1)]^{1-2(u_1 \oplus u_2 \oplus u_3 \oplus u_4)} LR(x_2) \quad LR(g_2) = [LR(x_3)]^{1-2(u_3 \oplus u_4)} LR(x_4)$$
$$LR(g_3) = [LR(x_5)]^{1-2(u_2 \oplus u_4)} LR(x_6) \quad LR(g_4) = [LR(x_7)]^{1-2u_4} LR(x_8). \qquad (2.44)$$

The likelihood ratios for h_1 and h_2 can be calculated as

$$LR(h_1) = [LR(g_1)]^{1-2(u_5 \oplus u_6)} LR(g_2) \quad LR(h_2) = [LR(g_3)]^{1-2u_6} LR(g_4). \qquad (2.45)$$

Fig. 2.33 Decoding path of u_8 when $N = 8$

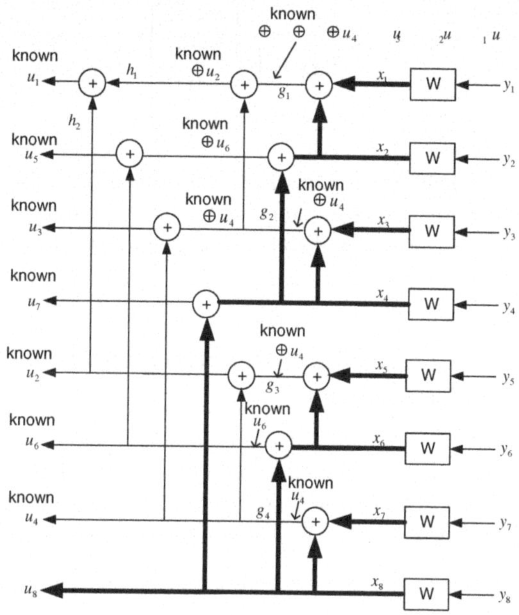

Fig. 2.34 Decoding tree of u_8 when $N = 8$

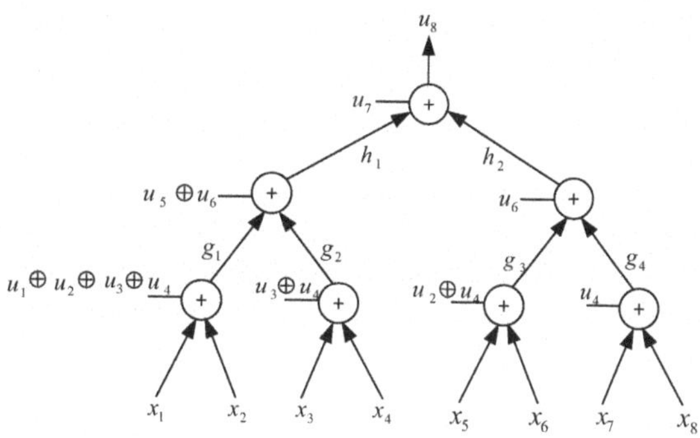

The likelihood ratio for information bit u_8 is calculated as

$$LR(u_8) = [LR(h_1)]^{1-2u_7} LR(h_2). \tag{2.46}$$

When the tree structure in Fig. 2.34 is inspected in details, we see that the node-bits can be formed by considering columns of the matrices G_4, G_2, G_1. The matrices are given below for reminder

$$G_4 = \begin{bmatrix} 1\,0\,0\,0 \\ 1\,0\,1\,0 \\ 1\,1\,0\,0 \\ 1\,1\,1\,1 \end{bmatrix} \quad G_2 = \begin{bmatrix} 1\,0 \\ 1\,1 \end{bmatrix} \quad G_1 = [1]. \tag{2.47}$$

It can be noticed that in the decoding operation of any data bit, some of the nodes at certain levels own some node bits in the tree structure. We will discuss in details the formation of node bits using generator matrices in the subsequent sections.

2.3 Level Indices and Determination of Levels for Bit Distribution

A tree for $N = 16$ is depicted in Fig. 2.35 where the levels of the tree are indicated by level-i. It is clear from Fig. 2.35 that the level with index i, that is level-i, has 2^i nodes. The total number of levels is $1 + \log_2 N$ where $N = 2^n$.

Example 2.2 For $N = 32$, how many levels are available in the decoding tree.

Solution 2.2 For $N = 32$ we have $1 + \log_2 32 = 6$ levels, and the indices of the levels are 0, 1, 2, 3, 4, 5. The topmost level has $2^0 = 1$ node and the bottommost level has $2^5 = 32$ nodes.

Determination of Nodes Which Owns Node Bits

Assume that L bits are decoded, and we want to decode the $(L + 1)^{\text{th}}$ bit. For this purpose we distribute the L previously decoded bits to the tree nodes. In this distribution stage, only the nodes at some levels get node bits. The index of the levels whose nodes get node bits can be determined from

$$L = \sum_i 2^i \tag{2.48}$$

where i denotes the index of the level whose nodes get node bits.

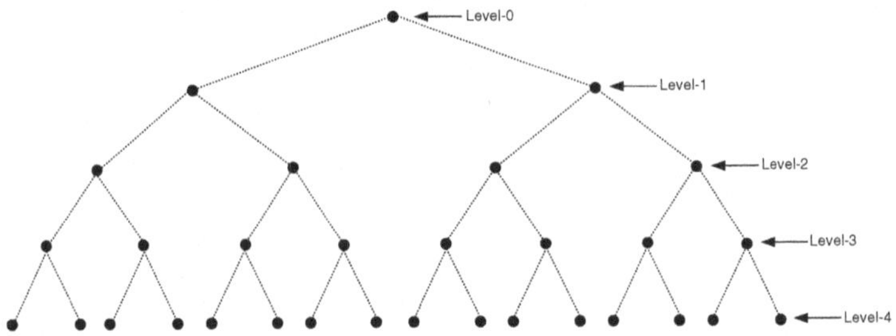

Fig. 2.35 Node levels when $N = 16$

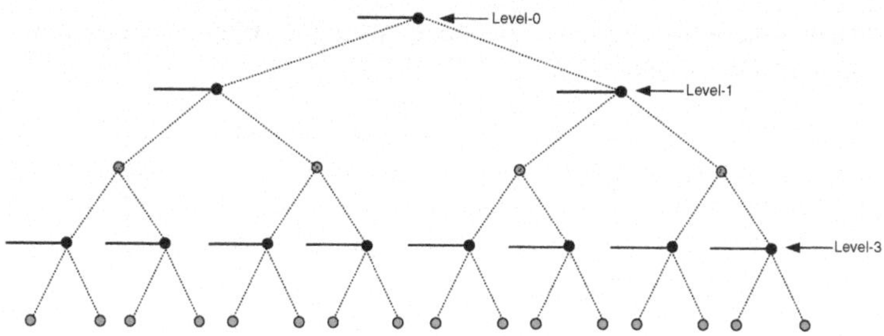

Fig. 2.36 Location of node bits for the decoding of 12th data bit

Example 2.3 Assume that $N = 16$, and we want to decode the 12th bit, i.e., the first 11 bits are decoded and they are to be distributed to the tree nodes. We can write 11 as

$$11 = 2^0 + 2^1 + 2^3$$

which implies that the nodes at levels with indices 0, 1, and 3 get node bits. This is illustrated in Fig. 2.36.

Exercise 2.4 *For $N = 16$, draw the tree structure for the decoding operation considering that*

(a) The first 6 information bits are decoded, and the 7th bit, i.e., u_7 is to be decoded.
(b) The first 10 information bits are decoded, and the 11th bit, i.e., u_{11} is to be decoded.
(c) The first 13 information bits are decoded, and the 14th bit, i.e., u_{14} is to be decoded.

Exercise 2.5 *Assume that $N = 1024$, and we want to decode the 867th bit. Determine the indices of the levels which get node bits after the distribution of the previously decoded 866 bits to the tree nodes.*

2.4 Decoding Algorithm for Polar Codes

The decoding approach detailed in the previous section can be expresses as a decoding algorithm which consists of two stages, and these stages are:

(a) Distribution of the previously decoded bits to the nodes.
(b) Decoding of the current bit.

The decoding tree consists of the repeated assembly of the tree-unit shown in Fig. 2.37 where the top node is called head node and left and right lower nodes are called left and right child nodes respectively.

The distribution of the previously decoded bits to the nodes can be achieved using the method explained in Algorithm 2.1.

Algorithm 2.1 Distribution of the decoded bits to the nodes:

Assume that i information bits are decoded, and we want to decode the $(i+1)$th information bit. The codeword vector has a length of N, i.e. the total number of bits to be decoded is N. We need to distribute the i decoded bits to the nodes of the decoding tree. This can be achieved as follows:

(1) Let $u_1^i = [u_1\, u_2 \ldots u_i]$
(2) If i is an odd number, then assign u_i as the node bit of the head-node. And calculate the left child-node and right node-child bits as

  (a) even bits $\rightarrow \bar{x}_e = (u_2\, u_4 \ldots u_{i-1})$
  (b) odd bits $\rightarrow \bar{x}_o = (u_1\, u_3 \ldots u_{i-2})$
  (c) left child – node bits $\rightarrow \bar{x}_l = \bar{x}_e \oplus \bar{x}_o$ where $\oplus$ denotes *XOR* operation
  (d) right child – node bits $\rightarrow \bar{x}_r = \bar{x}_e$

(3) $i = \lfloor i/2 \rfloor$, form the tree units for the left and right child node bits, and for the left child node

$$u_1^i = \bar{x}_l$$

and for the right child node

$$u_1^i = \bar{x}_r$$

and continue with step (2) for the left and right child nodes separately if $i \neq 1$, terminate otherwise.

The bit distribution stage outline in the presented algorithm can also be achieved using the columns of the generator matrices at certain levels. If a generator matrix contains M columns, then it can be used at the level which includes M nodes.

Example 2.4 The code word length for a polar code is $N = 8$. Assume that the first 5 bits are decoded as

$$u_1^5 = [1\,0\,1\,1\,1].$$

We want to decode the 6th bit, i.e., u_6. Distribute the previously decoded bits to the nodes so that decoding operation for the current bit can start.

Fig. 2.37 Tree unit

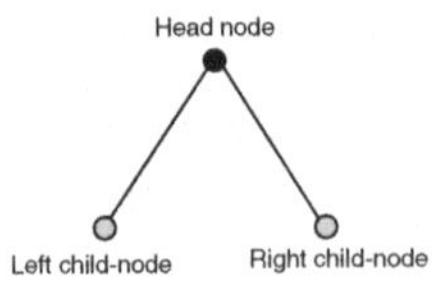

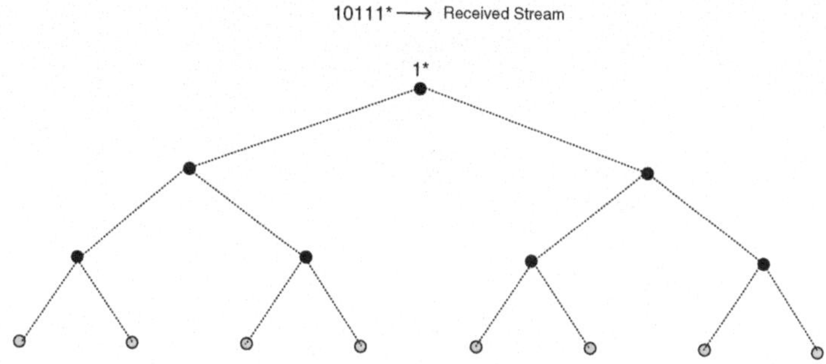

Fig. 2.38 Decoding tree with 5 decoded bits for the decoding of 6th bit

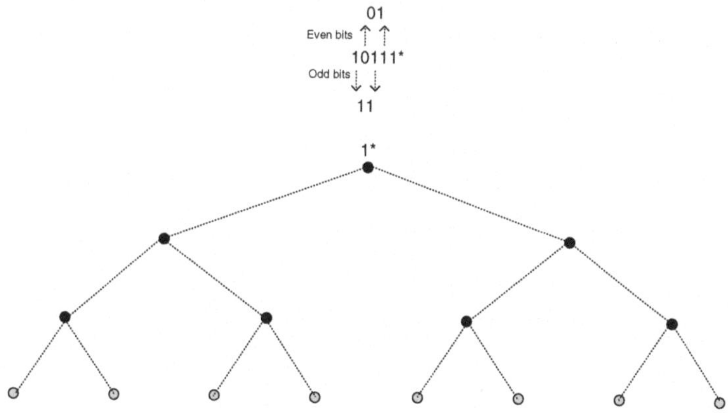

Fig. 2.39 First step of the distribution of decoded bits to the nodes

Solution 2.4 The decoded bit vector $u_1^5 = [u_1\, u_2\, u_3\, u_4\, u_5] = [1\,0\,1\,1\,1]$ contains an odd number of bits. For this reason, the decoded bit u_5 is assigned to topmost head node as shown in Fig. 2.38.

The remaining bits are grouped into two parts considering their even and odd indexes. This is illustrated in Fig. 2.39.

The even and odd bits are summed and assigned to the left node, and the even bits are assigned to the right node. This is explained in Fig. 2.40.

As it is seen from Fig. 2.40, the left and right node bits of the second level contains even number of bits. For both bit groups, we can construct even and odd groups and repeating the procedure in the previous step, we obtain the node-bits in Fig. 2.41.

At the end we get the node-bits as shown in Fig. 2.42, and we are ready for the second step of the decoding operation, i.e., combination of the likelihood ratios of the branches till the top most node and deciding on the value of the current bit.

The distribution of the previously decoded bits to the tree nodes can also be achieved using the generator matrices G_4 and G_1. Since $5 = 2^0 + 2^2$, the indices of the levels who get node bits are 0 and 2, and at these levels we have $2^0 = 1$ and $2^2 = 4$ nodes. The node bits at level-2 can be determined as

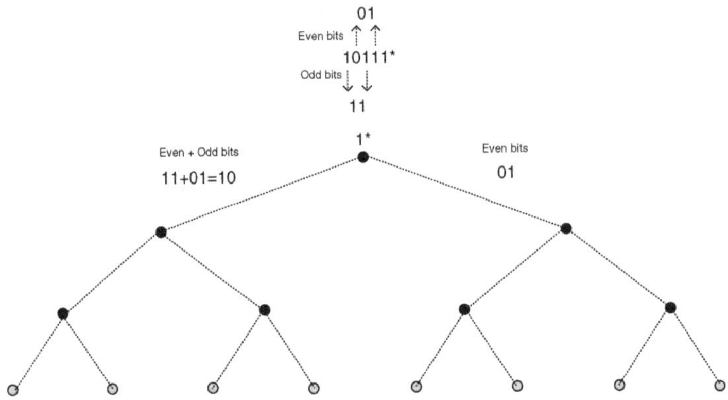

Fig. 2.40 Second step of the distribution of decoded bits to the nodes

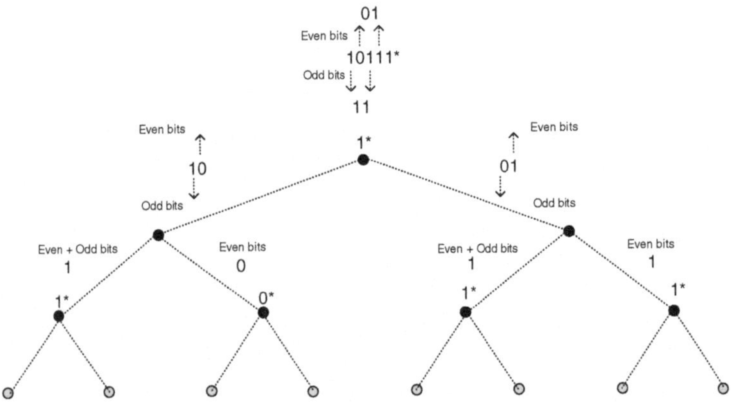

Fig. 2.41 Third step of the distribution of decoded bits to the nodes

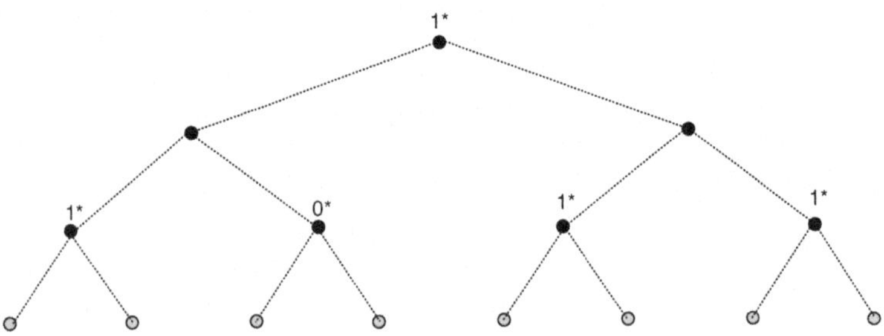

Fig. 2.42 Node bits

$$\bar{b}_2 = u_1^4 \times G_4 \rightarrow \bar{b}_2 = [1\ 0\ 1\ 1] \times \begin{bmatrix} 1\,0\,0\,0 \\ 1\,0\,1\,0 \\ 1\,1\,0\,0 \\ 1\,1\,1\,1 \end{bmatrix} \rightarrow \bar{b}_2 = [1\ 0\ 1\ 1] \quad (2.49)$$

When $\bar{b}_2 = [1\,0\,1\,1]$ is compared to the node bits at level-2 in Fig. 2.42 we see that they are the same. Since there is only a single note at level-0, the node bit equals to u_5 which is 1. In the next section, we study the distribution of previously decoded bits to the nodes using the generator matrices in details.

Definition 2.1 After bit distribution stage, if a node possesses a node bit, then that node is called g-node, otherwise, it is called f-node.

Distribution of the decoded bits to the nodes using the generator matrices

The decoded bits can be distributed to the nodes using the generator matrices. Assume that the code-word length is equal to N, and we decoded the first $M < N$ bits. The decoded bit vector can be written as

$$u_1^M = [u_1\, u_2 \ldots u_M].$$

For the decoding of $(M + 1)$th bit, we first distribute the previously decoded bits to the tree nodes. For this purpose, we first write M as sum of powers of 2, i.e., M is written as

$$M = \sum_i 2^i \quad (2.50)$$

where i indicates the indices of the levels where nodes are assigned bits. From (2.50), M can be written as

$$M = \underbrace{2^{i_1}}_{M_1} + \underbrace{2^{i_2}}_{M_2} + \cdots + \underbrace{2^{i_k}}_{M_k} \rightarrow M = M_1 + M_2 + \cdots + M_k.$$

In the second step, we divide the decoded bit vector u_1^M into sub-vectors $\bar{v}_1, \bar{v}_2, \cdots, \bar{v}_k$ containing $M_1, M_2, \ldots, M_k$ bits as

$$\left[\underbrace{u_1\, u_2 \ldots}_{\bar{v}_1} \ldots \underbrace{\ldots\ldots}_{\bar{v}k-1} \underbrace{\ldots u_{M_k}}_{\bar{v}_k} \right].$$

Once, we get the sub-vectors $\bar{v}_1, \bar{v}_2, \ldots, \bar{v}_k$, we can calculate the node bits at levels $i_1, i_2, \ldots, i_k$ as in (2.51)

$$\bar{b}_1 = \bar{v}_1 \times G_{M_1} \quad \bar{b}_2 = \bar{v}_2 \times G_{M_2} \ldots \bar{b}_k = \bar{v}_k \times G_{M_k}. \quad (2.51)$$

After obtaining the node bits for the levels $i_1, i_2, \ldots, i_k$, we can assign the bits to the corresponding nodes and start decoding of the current bit under concern.

Example 2.5 The code-word length for a polar code is $N = 16$. Assume that the first 7 bits are decoded as

$$u_1^7 = [u_1\, u_2\, u_3\, u_4\, u_5\, u_6\, u_7].$$

We want to decode the 8th bit, i.e., u_8. Distribute the previously decoded bits to the nodes so that decoding operation for the current bit can start. Use generator matrices for bit distribution operation.

Solution 2.5 First, we write the decoded bit number, i.e., 7, as sum of powers of two, i.e.

$$7 = \sum_i 2^i \rightarrow 7 = 2^0 + 2^1 + 2^2 \rightarrow 7 = 1 + 2 + 4.$$

Thus, we see that $7 = 1 + 2 + 4$. In the second step, we divide the decoded information sequence, starting from the last decoded bit towards to first decoded bit, into vectors $\bar{v}_1, \bar{v}_2, \bar{v}_4$ containing 1, 2, and 4 bits as illustrated in (2.52)

$$\left[\underbrace{u_1\, u_2\, u_3\, u_4}_{\bar{v}_4} \underbrace{u_5\, u_6}_{\bar{v}_2}\ \underbrace{u_7}_{\bar{v}_1} \right]. \tag{2.52}$$

Next, we multiply the vectors $\bar{v}_1, \bar{v}_2, \bar{v}_4$ by the generator matrices G_1, G_2, and G_4 and obtain the node bit vectors as

$$\bar{b}_1 = \bar{v}_1 \times G_1 \rightarrow \bar{b}_1 = u_7 \times 1 \rightarrow \bar{b}_1 = u_7$$

$$\bar{b}_2 = \bar{v}_2 \times G_2 \rightarrow \bar{b}_2 = [u_5\, u_6] \times \begin{bmatrix} 1\,0 \\ 1\,1 \end{bmatrix} \rightarrow \bar{b}_2 = [u_5 \oplus u_6 u_6]$$

$$\bar{b}_4 = \bar{v}_4 \times G_4 \rightarrow \bar{b}_4 = [u_1\, u_2\, u_3\, u_4] \times \begin{bmatrix} 1\,0\,0\,0 \\ 1\,0\,1\,0 \\ 1\,1\,0\,0 \\ 1\,1\,1\,1 \end{bmatrix} \rightarrow$$

$$\bar{b}_4 = [u_1 \oplus u_2 \oplus u_3 \oplus u_4 \quad u_3 \oplus u_4 \quad u_2 \oplus u_4 \quad u_4].$$

The bit in vector $\bar{b}_1$ is used for the node at level 0, and similarly the bits at the vectors $\bar{b}_2, \bar{b}_4$ are used for the nodes at the levels 1 and 2 respectively, i.e., powers of 2 in (2.50). When the calculated node bits are assigned to nodes at the corresponding levels, the decoding tree happens to be as in Fig. 2.43.

Example 2.6 The code word length for a polar code is $N = 16$. Assume that the first 12 bits are decoded as

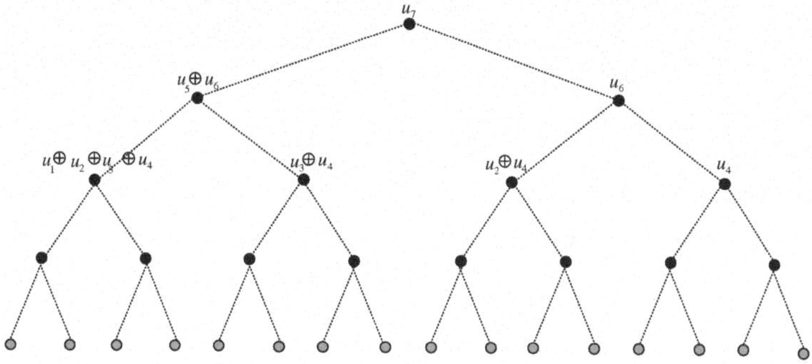

Fig. 2.43 Distribution of the decoded bits to the nodes using generator matrices

$$u_1^{12} = [1\,0\,1\,0\,0\,1\,0\,1\,1\,0\,1\,1].$$

We want to decode the 13th bit, i.e., u_{13}. Distribute the previously decoded bits to the nodes so that decoding operation for the current bit can start. Decide the levels where nodes are assigned to bits. Use generator matrices for bit distribution operation.

Solution 2.6 The number of decoded bits 12 can be written using powers of 2 as $12 = 2^2 + 2^3$. The decoded bit vector us divided into sub-vectors containing 4 and 8 bits as

$$\left[\underbrace{1\,0\,1\,0\,0\,1\,0\,1}_{\bar{v}_8}\underbrace{1\,0\,1\,1}_{\bar{v}_4}\right] \rightarrow \quad \bar{v}_4 = [1\,0\,1\,1] \quad \bar{v}_8 = [1\,0\,1\,0\,0\,1\,0\,1].$$

The node bits at level 2 and at level 3 are calculated as

$$\bar{b}_4 = \bar{v}_4 \times G_4 \rightarrow \bar{b}_4 = [1\,0\,1\,1] \times \begin{bmatrix} 1\,0\,0\,0 \\ 1\,0\,1\,0 \\ 1\,1\,0\,0 \\ 1\,1\,1\,1 \end{bmatrix} \rightarrow \bar{b}_4 = [1\,0\,1\,1]$$

$$\bar{b}_8 = \bar{v}_8 \times G_8 \rightarrow$$

$$\bar{b}_8 = [1\,0\,1\,0\,0\,1\,0\,1] \times \begin{bmatrix} 1\,0\,0\,0\,0\,0\,0\,0 \\ 1\,0\,0\,0\,1\,0\,0\,0 \\ 1\,0\,1\,0\,0\,0\,0\,0 \\ 1\,0\,1\,0\,1\,0\,1\,0 \\ 1\,1\,0\,0\,0\,0\,0\,0 \\ 1\,1\,0\,0\,1\,1\,0\,0 \\ 1\,1\,1\,1\,0\,0\,0\,0 \\ 1\,1\,1\,1\,1\,1\,1\,1 \end{bmatrix} \rightarrow \bar{b}_8 = [0\,0\,0\,1\,0\,0\,1\,1].$$

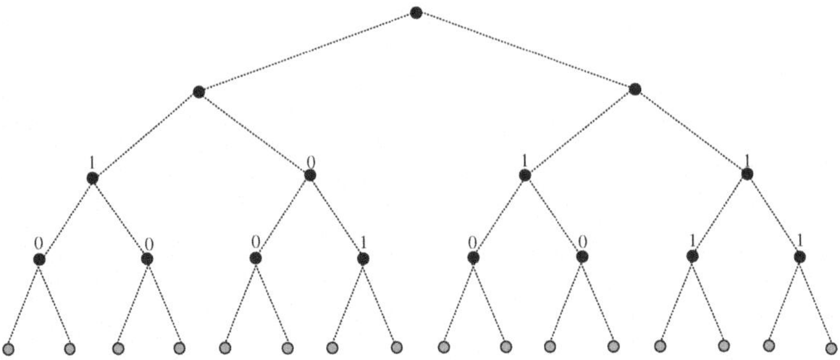

Fig. 2.44 Determination of nodes for the decoding of 12th bit

Fig. 2.45 Decoder kernel unit

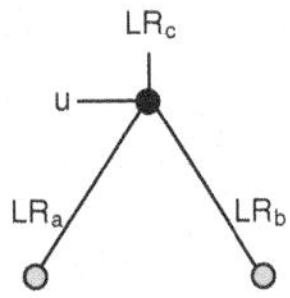

When the calculated node bits are assigned to the nodes at the levels 0 and 3, the decoding tree happens to be as in Fig. 2.44.

Exercise 2.7 *The code word length for a polar code is* $N = 32$. *Assume that the first 21 bits are decoded as*

$$u_1^{12} = [1\,0\,1\,0\,0\,1\,0\,1\,1\,0\,1\,1\,1\,1\,0\,1\,0\,0\,0\,1\,1].$$

We want to decode the 22nd bit, i.e., u_{22}. *Distribute the previously decoded bits to the nodes so that decoding operation for the current bit can start. Decide the levels where nodes are assigned to bits. Use generator matrices for bit distribution operation.*

Algorithm 2.2 Decoding of the current bit:

Once the previously bits are distributed to the nodes, in the next step, decoding operation takes place. Decoding operation starts from the lowest level and goes up to the top most head node. From bottom to top, at every node, likelihood ratio calculation is performed, and at the top most head node considering the calculated likelihood ratio the bit value is decided.

Consider the tree unit shown in Fig. 2.45. If the head node has a node bit u, then the likelihood ratio of the head node is calculated

$$LR_c = LR_a^{1-2u} LR_b \rightarrow LR_c = LR_a LR_b \quad \text{for}\, u = 0, \quad \text{and} \quad LR_c = \frac{LR_b}{LR_a} \quad \text{for}\, u = 1. \quad (2.53)$$

Example 2.8 Calculate the likelihood ratio of the top most head node for the tree shown in Fig. 2.46.

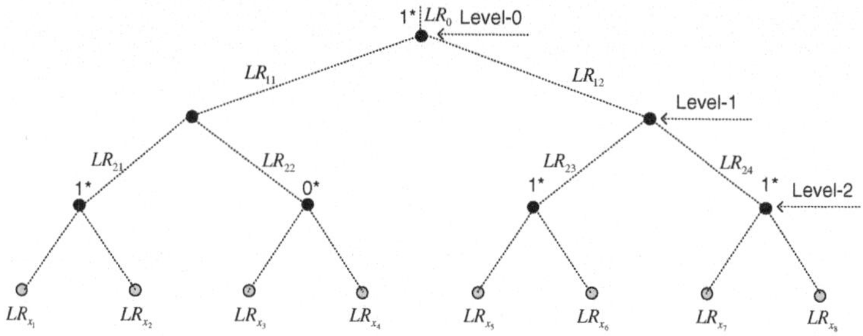

Fig. 2.46 Decoding tree for the Example 2.8

Solution 2.8 The likelihood ratios $LR_{21}, LR_{22}, LR_{23}, LR_{24}$ are calculated using the formula,

$$LR_{head} = \left(LR_{left}\right)^{1-2u} LR_{right} \tag{2.54}$$

where u is the node bit, as

$$\begin{aligned} LR_{21} &= \left(LR_{x_1}\right)^{-1} LR_{x_2} \quad LR_{22} = LR_{x_3} LR_{x_4} \\ LR_{23} &= \left(LR_{x_5}\right)^{-1} LR_{x_6} \quad LR_{24} = \left(LR_{x_7}\right)^{-1} LR_{x_8} \end{aligned} \tag{2.55}$$

The likelihoods LR_{21}, LR_{22} can be calculated using the formula

$$LR_{head} = \frac{1 + LR_{left} LR_{right}}{LR_{left} + LR_{right}} \tag{2.56}$$

as

$$LR_{11} = \frac{1 + LR_{21} LR_{22}}{LR_{21} + LR_{22}} \quad LR_{12} = \frac{1 + LR_{23} LR_{24}}{LR_{23} + LR_{24}}. \tag{2.57}$$

Finally, the likelihood LR_{01} is calculated as

$$LR_0 = (LR_{11})^{-1} LR_{12}. \tag{2.58}$$

2.5 Binary Erasure Channel (BEC)

The binary erasure channel is depicted in Fig. 2.47 where α is the erasure probability of the channel and it is defined as $\alpha = Prob(y = e|x = 0)$ or $\alpha = Prob(y = e|x = 1)$.

The mutual information between channel input and channel output is given as

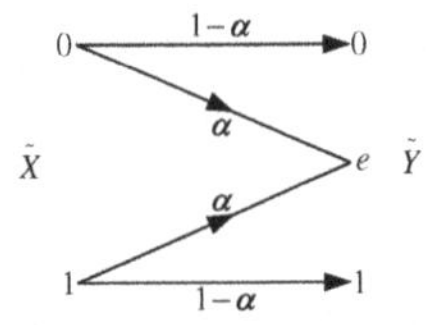

Fig. 2.47 Binary erasure channel

$$I\left(\tilde{X};\tilde{Y}\right) = H\left(\tilde{X}\right)(1-\alpha) \tag{2.59}$$

where $H(\tilde{X})$ is the entropy of the channel input and is calculated as

$$H\left(\tilde{X}\right) = \sum_x p(x)\log\frac{1}{p(x)}. \tag{2.60}$$

The capacity of the binary erasure channel can be calculated using

$$C = \max_{p(x)} I\left(\tilde{X};\tilde{Y}\right) \tag{2.61}$$

as

$$C = 1-\alpha. \tag{2.62}$$

Example 2.9 Assume that at the output of the binary erasure channel we have the symbol y, and input symbols are equally likely, i.e., $Prob(x=0) = Prob(x=1) = 0.5$. Then, for $y=0$, $y=1$, and $y=-1$, the probability ratio

$$LR = \frac{Prob(x=0|y)}{Prob(x=1|y)}$$

takes the values

$$LR_0 = \frac{Prob(x=0|y=0)}{Prob(x=1|y=0)} \to LR_0 = \frac{Prob(y=0|x=0)Prob(x=0)}{Prob(y=0|x=1)Prob(x=1)} \to$$
$$LR_0 = \frac{1-\alpha}{0} \to LR_0 = \infty.$$
$$LR_1 = \frac{Prob(x=0|y=1)}{Prob(x=1|y=1)} \to LR_1 = \frac{Prob(y=1|x=0)Prob(x=0)}{Prob(y=1|x=1)Prob(x=1)} \to$$
$$LR_1 = \frac{0}{1-\alpha} \to LR_1 = 0.$$
$$LR_{-1} = \frac{Prob(x=0|y=-1)}{Prob(x=1|y=-1)} \to LR_{-1} = \frac{Prob(y=-1|x=0)Prob(x=0)}{Prob(y=-1|x=1)Prob(x=1)} \to$$
$$LR_{-1} = \frac{\alpha}{\alpha} \to L_{-1} = 1.$$

Table 2.1 Likelihood ratio table

LR_a	LR_b	LR_c	LR_d	LR_e
0	1	1	0	100
0	100	0.01	0	100
1	0	1	0	0
1	100	1	100	100
100	0	0.01	0	0
100	1	1	100	0.01

In practical implementations, LR_0 can be taken as any large number, for instance $LR_0 = 100$.

Example 2.10 Assume that the likelihoods LR_a and LR_b can take a value from the set $\{0, 1, 100\}$. Calculate the possible values of following likelihoods

$$LR_c = \frac{1 + LR_a LR_b}{LR_a + LR_b} \quad LR_d = LR_a LR_b \quad LR_e = \frac{LR_b}{LR_a}. \tag{2.63}$$

Solution 2.10 The possible values of LR_c, LR_d, and LR_e are depicted in Table 2.1 where we used 100 for the place of ∞.

Example 2.11 Consider the 4-bit polar encoder depicted in Fig. 2.8. Assume that binary erasure channels are used for the transmission of code bits and the channel outputs are

$$y_1 = 1 \quad y_2 = 0 \quad y_3 = -1 \quad y_4 = 1.$$

Calculate the likelihood ratios for the code bits x_i, $i = 1, \ldots, 4$.

Solution The likelihood ratios for the code bits can be calculated are defined as

$$LR(x_1) = \frac{P(y_1|x_1 = 0)}{P(y_1|x_1 = 1)} \quad LR(x_2) = \frac{P(y_2|x_2 = 0)}{P(y_2|x_2 = 1)}$$
$$LR(x_3) = \frac{P(y_3|x_3 = 0)}{P(y_3|x_3 = 1)} \quad LR(x_4) = \frac{P(y_4|x_4 = 0)}{P(y_4|x_4 = 1)}$$

which can be calculated using the binary erasure channel transition probabilities as

$$LR(x_1) = \frac{P(y_1 = 1|x_1 = 0)}{P(y_1 = 1|x_1 = 1)} \rightarrow LR(x_1) = 0 \quad LR(x_2) = \frac{P(y_2 = 0|x_2 = 0)}{P(y_2 = 0|x_2 = 1)} \rightarrow LR(x_2) = 100$$
$$LR(x_3) = \frac{P(y_3 = -1|x_3 = 0)}{P(y_3 = -1|x_3 = 1)} \rightarrow LR(x_3) = 1 \quad LR(x_4) = \frac{P(y_4 = 1|x_4 = 0)}{P(y_4 = 1|x_4 = 1)} \rightarrow LR(x_4) = 0.$$

2.6 Determination of Frozen Bit Locations for BEC Channels

The Bhattacharyya parameters of the split binary erasure channels can be computed in a recursive manner as in

$$\begin{aligned} Z\left(W_N^{2k-1}\right) &= 2Z\left(W_{N/2}^k\right) - \left[Z\left(W_{N/2}^k\right)\right]^2 \\ Z\left(W_N^{2k}\right) &= \left[Z\left(W_{N/2}^k\right)\right]^2 \end{aligned} \tag{2.64}$$

in which the initial value for the recursion is chosen as

$$Z\left(W_1^1\right) = \alpha$$

where α is the erasure probability of the binary erasure channel. And for the binary erasure split channels, the channel capacities can be calculated from the Bhattacharyya parameters as in

$$I\left(W_N^k\right) = 1 - Z\left(W_N^k\right). \tag{2.65}$$

Note that Bhattacharyya parameter can be considered as the maximum value of the probability of transmission error.

Example 2.12 Let $N = 4$ and $\alpha = 0.5$, the Bhattacharyya parameters for the split channels can be calculated as follows.

For $N = 1$, (2.64) reduces to

$$Z\left(W_1^1\right) = 0.5$$

For $N = 2$, (2.64) reduces to

$$\begin{aligned} Z\left(W_2^{2k-1}\right) &= 2Z\left(W_1^k\right) - \left[Z\left(W_1^k\right)\right]^2 \\ Z\left(W_2^{2k}\right) &= \left[Z\left(W_1^k\right)\right]^2 \end{aligned}$$

which can be evaluated for $k = 1$ as

$$\begin{aligned} &Z\left(W_2^1\right) = 2Z\left(W_1^1\right) - \left[Z\left(W_1^1\right)\right]^2 \rightarrow Z\left(W_2^1\right) = 2 \times 0.5 - [0.5]^2 \rightarrow Z\left(W_2^1\right) = 0.75 \\ &Z\left(W_2^2\right) = \left[Z\left(W_1^1\right)\right]^2 \rightarrow Z\left(W_2^2\right) = [0.5]^2 \rightarrow Z\left(W_2^2\right) = 0.25. \end{aligned}$$

For $N = 4$, (2.64) reduces to

$$\begin{aligned} Z\left(W_4^{2k-1}\right) &= 2Z\left(W_2^k\right) - \left[Z\left(W_2^k\right)\right]^2 \\ Z\left(W_4^{2k}\right) &= \left[Z\left(W_2^k\right)\right]^2 \end{aligned}$$

which can be evaluated for $k = 1$ as

$$Z\left(W_4^1\right) = 2Z\left(W_2^1\right) - \left[Z\left(W_2^1\right)\right]^2 \rightarrow Z\left(W_4^1\right) = 2 \times 0.75 - [0.75]^2 \rightarrow Z\left(W_4^1\right) = 0.9375$$
$$Z\left(W_4^2\right) = \left[Z\left(W_2^1\right)\right]^2 \rightarrow Z\left(W_4^2\right) = [0.75]^2 \rightarrow Z\left(W_4^2\right) = 0.5625$$

and for $k = 2$ as

$$Z\left(W_4^3\right) = 2Z\left(W_2^2\right) - \left[Z\left(W_2^2\right)\right]^2 \rightarrow Z\left(W_4^3\right) = 2 \times 0.25 - 0.25^2 \rightarrow Z\left(W_4^3\right) = 0.4375$$
$$Z\left(W_4^4\right) = \left[Z\left(W_2^2\right)\right]^2 \rightarrow Z\left(W_4^4\right) = 0.0625.$$

Thus, the split channel capacities can be written in a vector as

$$Z\left(W_4^i\right) = [0.9375\ 0.5625\ 0.4375\ 0.0625] \tag{2.66}$$

which implies that the split channel W_4^4 has the least probability error, and the split channel W_4^1 has the largest probability of error.

If we choose the code rate as 0.5, then 2 information and 2 frozen bits will be utilized for the transmission. Considering the split channel Bhattacharyya parameters evaluated in (2.66), we construct the frame to be encoded by the polar encoder as

$$\bar{u} = [0\,0\,u_3\,u_4]$$

and when encoding operation is performed, we obtain the code-word as

$$x_1^4 = \bar{u}G_4 \rightarrow x_1^4 = [0\,0\,u_3\,u_4] \times \begin{bmatrix} 1\,0\,0\,0 \\ 1\,0\,1\,0 \\ 1\,1\,0\,0 \\ 1\,1\,1\,1 \end{bmatrix} \rightarrow x_1^4 = [u_3 \oplus u_4 \quad u_3 \oplus u_4 \quad u_4 \quad u_4].$$

In addition, considering the indices of the Bhattacharyya values of the split channels in (2.66), we can sort the capacities of the split channels in decreasing order as

$$\bar{C} = [4\,3\,2\,1]$$

which implies that the split channel W_4^4 has the largest capacity and the split channel W_4^1 has the smallest capacity.

Example 2.13 Sort the split channel capacities in ascending order for a polar encoder with code-word length $N = 8$, and for code rate $R = 4/8$, obtain the code-word bits in terms of the data bits.

Solution 2.13 When the recursive equations in (2.64) are solved for $N = 8$, we obtain the Bhattacharyya values of the split channels as in

PR 2.1 Calculation of the Bhattacharyya parameters

```
function [Z_Unsorted,Z_Sorted,C_Index]=compute_Z_Params(n,e)
% e is the erasure probability

N=2^n;
currN=2.^[1:n];
Z(1,1)=e;

for j=1:N/2
  for indx=1:n
    Z(currN(indx),2*j-1)=2*Z(currN(indx)/2,j)-(Z(currN(indx)/2,j))^2;
    Z(currN(indx),2*j)=(Z(currN(indx)/2,j))^2;
  end
end

Z_Unsorted=Z(N,:);
[Z_Sorted,C_Index] = sort(Z(N,:));
```

$$Z\left(W_8^i\right) = [0.9961 \quad 0.8789 \quad 0.8086 \quad 0.3164 \quad 0.6836 \quad 0.1914 \quad 0.1211 \quad 0.0039]$$

which can be used to sort the capacities of the split channels using their indices in ascending order as

$$\bar{C} = [8 \quad 7 \quad 6 \quad 4 \quad 5 \quad 3 \quad 2 \quad 1].$$

For the code rate $R = 4/8$, we can form the data frame to be encoded as

$$\bar{u} = [0\,0\,0\,u_1\,0\,u_2\,u_3\,u_4]$$

and the code-word can be generated as

$$x_1^8 = \bar{u}G_8 \rightarrow x_1^8 = [0\,0\,0\,u_1\,0\,u_2\,u_3\,u_4]\begin{bmatrix} 1\,0\,0\,0\,0\,0\,0\,0 \\ 1\,0\,0\,0\,1\,0\,0\,0 \\ 1\,0\,1\,0\,0\,0\,0\,0 \\ 1\,0\,1\,0\,1\,0\,1\,0 \\ 1\,1\,0\,0\,0\,0\,0\,0 \\ 1\,1\,0\,0\,1\,1\,0\,0 \\ 1\,1\,1\,1\,0\,0\,0\,0 \\ 1\,1\,1\,1\,1\,1\,1\,1 \end{bmatrix}$$

resulting in

$$x_1^8 = [u_1 \oplus u_2 \oplus u_3 \oplus u_4 \quad u_2 \oplus u_3 \oplus u_4 \quad u_1 \oplus u_3 \oplus u_4 \quad u_3 \oplus u_4 \quad u_1 \oplus u_2 \oplus u_4 \quad u_2 \oplus u_4 \quad u_1 \oplus u_4 \quad u_4].$$

Example 2.14 Write a Matlab program for the calculation of the Bhattacharyya parameters of the polar encoders utilizing binary erasure channels with erasure probability α.

Solution 2.14 A Matlab function as in PR 2.1 can be written to calculate the Bhattacharyya parameters and to sort the capacities of the split channels in descending order for polar encoders employing binary erasure channels.

Second approach for the calculation of frozen bit locations for polar encoders employing BECs

Another method for the calculation of frozen bit locations similar to the approach in (2.64) is introduced in [3]. However, in this approach the generator matrix is defined as

$$G'_N = F^{\otimes n}.$$

On the other, for the approach in (2.64), the generator matrix is defined as

$$G_N = B_N F^{\otimes n}.$$

Assume that $P(\cdot)$ is the column permutation function such that

$$G_N = P(G'_N).$$

If $\bar{C}'$, and $\bar{C}$ are the vector of split channel indices in ascending order such that G'_N is used for $\bar{C}'$ and G_N is used for $\bar{C}$, then we have

$$\bar{C} = P(\bar{C}').$$

At the end of Chap. 1, we discussed the difference and similarities between G_N and G'_N. We suggest to the reader to review the last section of Chap. 1.

Let's now explain the method in [3]. To find the location of frozen bits, first for a block length of $N = 2^n, n \geq 1$, the row vector

$$z_N = [z(N, 1)z(N, 2)\ldots z(N, N)] \tag{2.67}$$

is computed using [3]

$$z(2k, j) = \begin{cases} 2z(k, j) - (z(k, j))^2 & for\ 1 \leq j \leq k \\ (z(k, j - k))^2 & for\ k + 1 \leq j \leq 2k \end{cases} \tag{2.68}$$

where $k = 1, 2, 2^2, \ldots, 2^{n-1}$ with the initial value $z(1, 1) = \alpha$ where α is the erasure probability of the BEC. In the next step, the permutation vector

$$\pi_N = [i_1 i_2 \ldots i_N] \tag{2.69}$$

of the set $[12\ldots N]$ is formed considering the constraint

$$z(N, i_j) < z(N, i_k) \quad 1 \leq j < k \leq N. \tag{2.70}$$

Example 2.15 Calculate z_N and π_N for $N = 2^2$, take $\alpha = 1/2$.

Solution 2.15 For $k = 1$, (2.68) reduces to

$$z(2, j) = \begin{cases} 2z(1, j) - (z(1, j))^2 & \text{for } 1 \le j \le 1 \\ (z(1, j-1))^2 & \text{for } 2 \le j \le 2 \end{cases} \tag{2.71}$$

which can be calculated for $j = 1, j = 2$ using $z(1, 1) = 1/2$ as

$$z(2, 1) = 2z(1, 1) - (z(1, 1))^2 \rightarrow z(2, 1) = 1 - \frac{1}{4} \rightarrow z(2, 1) = \frac{3}{4}$$
$$z(2, 2) = (z(1, 1))^2 \rightarrow z(2, 2) = \frac{1}{4}$$

For $k = 2$, (2.68) reduces to

$$z(4, j) = \begin{cases} 2z(2, j) - (z(2, j))^2 & \text{for } 1 \le j \le 2 \\ (z(2, j-2))^2 & \text{for } 3 \le j \le 4 \end{cases} \tag{2.72}$$

which can be calculated for $j = 1, \cdots, 4$ using $z(1, 1) = 1/2, z(2, 1) = 3/4$, $z(2, 2) = 1/4$ as

$$z(4, 1) = 2z(2, 1) - (z(2, 1))^2 \rightarrow z(4, 1) = \frac{3}{2} - \frac{9}{16} \rightarrow z(4, 1) = \frac{15}{16}$$
$$z(4, 2) = 2z(2, 2) - (z(2, 2))^2 \rightarrow z(4, 2) = \frac{1}{2} - \frac{1}{16} \rightarrow z(4, 2) = \frac{7}{16}$$
$$z(4, 3) = (z(2, 1))^2 \rightarrow z(4, 3) = \frac{9}{16}$$
$$z(4, 4) = (z(2, 2))^2 \rightarrow z(4, 4) = \frac{1}{16}.$$

Using the calculated values, we can form the vector

$$z_4 = [z(4, 1) \quad z(4, 2) \quad z(4, 3) \quad z(4, 4)]$$

as

$$z_4 = \left[\frac{15}{16} \frac{7}{16} \frac{9}{16} \frac{1}{16}\right]$$

To construct π_4, we pay attention to the elements of z_4, and sort the indices of the elements considering the magnitudes in ascending order. That is:

$$z_4 = \left[\underbrace{\frac{15}{16}}_{1} \underbrace{\frac{7}{16}}_{2} \underbrace{\frac{9}{16}}_{3} \underbrace{\frac{1}{16}}_{4}\right] \rightarrow \pi_4 = [4\,2\,3\,1] \tag{2.73}$$

The channel capacity is related to the elements of z_N in an opposite manner, i.e., if the element of z_N is large, the corresponding capacity is low and vice versa, and the elements of π_N are the orders of the capacities in descending manner. That is, if π_4 is considered we can order the channel capacities as

$$C_4 > C_2 > C_3 > C_1.$$

That is, the bit transmitted at the 4th place passes through the channel having the largest capacity, i.e., the most reliable channel, and the bit transmitted at the first place passes through the channel having the smallest capacity.

This means that, if we want to use some frozen bits, i.e., parity bits, then we should choose the frozen bits, i.e., 0 bits, considering the lowest capacity channels first and choose relatively greater one if low rate transmission is allowed.

Considering π_4, if we use only one frozen bit, then the transmitted sequence becomes as

$$u = [0\,d\,d\,d]$$

and the rate of the transmission is 3/4. If we use two frozen bits, then the transmitted sequence would be as

$$u = [0\,d\,0\,d]$$

and the rate of the transmission is 2/4. If we use three frozen bits, then the transmitted sequence would be as

$$u = [0\,0\,0\,d]$$

and the rate of the transmission is 1/4.

A computer program can be written for the calculation of z_N and π_4 for medium or large N values.

Example 2.16 Determine the locations of the frozen bits for a transmitted sequence of length $N = 2^3$ for a transmission rate of $R = 4/8$ for binary erasure channel with erasure probability $\alpha = 1/2$.

Solution 2.16 The vector z_N in Eq. (1.10a) can be calculated for $N = 2^3$ writing a computer program, such as writing a Matlab script, as in

$$z_8 = [0.9961 \quad 0.6836 \quad 0.8086 \quad 0.1211 \quad 0.8789 \quad 0.1914 \quad 0.3164 \quad 0.0039].$$

and π_8 vector can be written as

$$\pi_8 = [8 \quad 4 \quad 6 \quad 7 \quad 2 \quad 3 \quad 5 \quad 1].$$

On the other hand, if (2.64) is used for the calculation of channel indices, then we would obtain the capacity indices vector as

$$\bar{C} = [8 \quad 7 \quad 6 \quad 4 \quad 5 \quad 3 \quad 2 \quad 1].$$

Considering π_8, the capacities of channels can be sorted as

$$C_8 > C_4 > C_6 > C_7 > C_2 > C_3 > C_5 > C_1$$

which can be written in increasing order as a vector as

$$\bar{C} = [C_1\, C_5\, C_3\, C_2\, C_7\, C_6\, C_4\, C_8].$$

This means that the bit transmitted at the 8th place passes through the channel having the largest capacity, and the bit transmitted at the 1st place passes through the channel having the smallest capacity. Since the frozen bits are located to the lowest capacity channels first, for rate $R = 4/8$, considering the capacity order, the transmitted sequence can take the form

$$u = [0\,0\,0\,d\,0\,d\,d\,d].$$

Example 2.17 Calculate z_N, π_N and $\bar{C}$ for $N = 2^4$, take $\alpha = 1/2$.

Solution 2.17 z_{16}, π_{16} and $\bar{C}$ can be found as

$$\begin{aligned} z_{16} = [&1.0000 \quad 0.8999 \quad 0.9634 \quad 0.2275 \quad 0.9853 \quad 0.3462 \quad 0.5327 \quad 0.0078 \quad 0.9922 \\ &0.4673 \quad 0.6538 \quad 0.0147 \quad 0.7725 \quad 0.0366 \quad 0.1001 \quad 0.0000] \\ \pi_{16} = [&16 \quad 8 \quad 12 \quad 14 \quad 15 \quad 4 \quad 6 \quad 10 \quad 7 \quad 11 \quad 13 \quad 2 \quad 3 \quad 5 \quad 9 \quad 1] \\ \bar{C} = [&C_1 \quad C_5 \quad C_9 \quad C_3 \quad C_2 \quad C_{13} \quad C_{11} \quad C_7 \quad C_{10} \quad C_6 \quad C_4 \quad C_{15} \quad C_{14} \quad C_{12} \quad C_8 \quad C_{16}]. \end{aligned}$$

Let's now consider a complete example involving BEC and successive cancellation decoding of polar codes.

2.7 Decoding Operation with Frozen Bits

In this section we explain the decoding of polar codes using a complete example. In this example, the channel capacities calculated for binary erasure channel with erasure probability $\alpha = 0.5$ and $N = 8$ are used, i.e., we use the capacity sorting vector

$$\pi_8 = [8 \quad 7 \quad 6 \quad 4 \quad 5 \quad 3 \quad 2 \quad 1].$$

Example 2.18 Let the data length and rate be as $N = 8$, and $R = 0.5$ respectively. Choose the information bit vector as

$$\bar{b} = [1\,1\,1\,1].$$

Considering the locations of frozen bits, the data vector to be transmitted is formed seeding the frozen bits to the proper locations as

$$\bar{d} = \begin{bmatrix}0^* & 0^* & 0^* & 1 & 0^* & 1 & 1 & 1\end{bmatrix}$$

where 0^* indicates the frozen bits, i.e., parity bits used for error correction purposes.

The encoding operation is performed using the generator matrix

$$G = \begin{bmatrix} 1\,0\,0\,0\,0\,0\,0\,0 \\ 1\,0\,0\,0\,1\,0\,0\,0 \\ 1\,0\,1\,0\,0\,0\,0\,0 \\ 1\,0\,1\,0\,1\,0\,1\,0 \\ 1\,1\,0\,0\,0\,0\,0\,0 \\ 1\,1\,0\,0\,1\,1\,0\,0 \\ 1\,1\,1\,1\,0\,0\,0\,0 \\ 1\,1\,1\,1\,1\,1\,1\,1 \end{bmatrix}$$

and data vector $\bar{d} = [0^* \quad 0^* \quad 0^* \quad 1 \quad 0^* \quad 1 \quad 1 \quad 1]$ as

$$\bar{x} = \bar{d}G \rightarrow \bar{x} = [0 \quad 0 \quad 0 \quad 1 \quad 0 \quad 1 \quad 1 \quad 1] \begin{bmatrix} 1\,0\,0\,0\,0\,0\,0\,0 \\ 1\,0\,0\,0\,1\,0\,0\,0 \\ 1\,0\,1\,0\,0\,0\,0\,0 \\ 1\,0\,1\,0\,1\,0\,1\,0 \\ 1\,1\,0\,0\,0\,0\,0\,0 \\ 1\,1\,0\,0\,1\,1\,0\,0 \\ 1\,1\,1\,1\,0\,0\,0\,0 \\ 1\,1\,1\,1\,1\,1\,1\,1 \end{bmatrix}$$

leading to the code-word

$$\bar{x} = [0 \quad 1 \quad 1 \quad 0 \quad 1 \quad 0 \quad 0 \quad 1].$$

Let's transmit the vector $\bar{x}$ through the binary erasure channel with erasure probability is $\alpha = 0.5$. Assume the channel output is

$$\bar{y} = [0 \quad 1 \quad 1 \quad -1 \quad -1 \quad 0 \quad 0 \quad -1]$$

where - 1's indicate the erased symbols.

Decoding of first information bit, i.e., u_4:

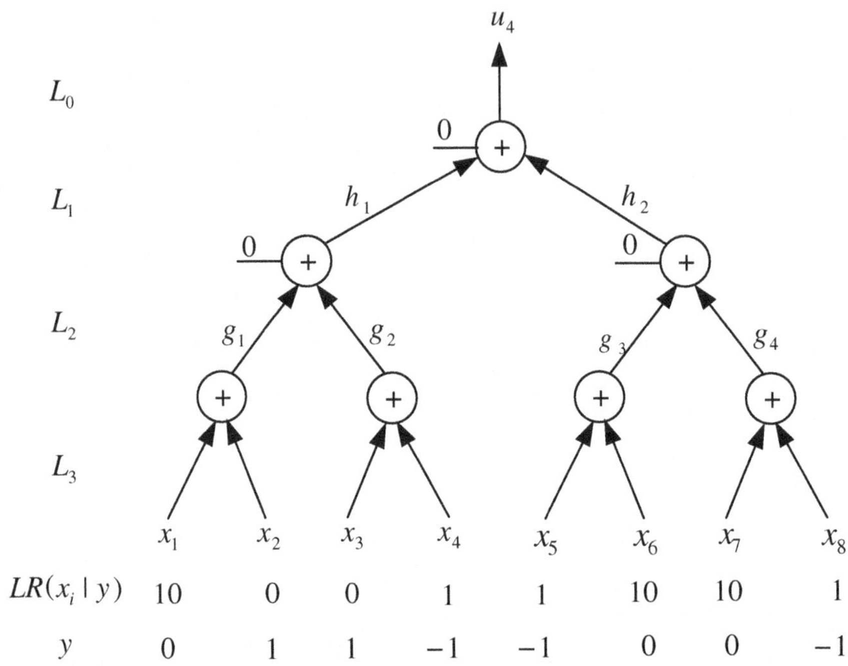

Fig. 2.48 Decoding tree for data bit u_4

Using the received signal vector $\bar{y}$, we can calculate the likelihood ratios for bottom most level as

$$LR(x_1|y_1) = 10 \quad LR(x_2|y_2) = 0 \quad LRx_3|y_3 = 0 \quad LR(x_4|y_4) = 1$$
$$LR(x_5|y_5) = 1 \quad LRx_6|y_6 = 10 \quad LR(x_7|y_7) = 10 \quad LR(x_8|y_8) = 1$$

which can be written as likelihood vector as

$$L_3 = [10 \quad 0 \quad 0 \quad 1 \quad 1 \quad 10 \quad 10 \quad 1].$$

Since the first 3-bits are frozen bits, i.e., 0's, the first 3 decoded bits can be written as a vector

$$\bar{u} = [u_1 = 0\, u_2 = 0\, u_3 = 0].$$

We can distribute the first decoded 3-bits to the nodes as in Fig. 2.48. The decoding tree and likelihood vector L_3 is depicted in Fig. 2.48.

We can now decode the data bit u_4.

Decoding of u_4:

The likelihoods in level-2, i.e., LR_2, can be calculated as

$$LR(g_1) = \frac{1 + LR(x_1)LR(x_2)}{LR(x_1) + LR(x_2)} \quad LR(g_2) = \frac{1 + LR(x_3)LR(x_4)}{LR(x_3) + LR(x_4)}$$
$$LR(g_3) = \frac{1 + LR(x_5)LR(x_6)}{LR(x_5) + LR(x_6)} \quad LR(g_4) = \frac{1 + LR(x_7)LR(x_8)}{LR(x_7) + LR(x_8)} \tag{2.74}$$

leading to

$$LR(g_1) = 0.1 \quad LR(g_2) = 1 \quad LR(g_3) = 1 \quad LR(g_4) = 1$$

which can be written as likelihood vector as

$$L_2 = [0.1 \quad 1 \quad 1 \quad 1].$$

The likelihoods in level-1, i.e., L_1, can be calculated as

$$LR(h_1) = (LR(g_1))^{1-2\times 0} LR(g_2) \rightarrow LR(h_1) = LR(g_1)LR(g_2)$$
$$LR(h_2) = (LR(g_3))^{1-2\times 0} LR(g_4) \rightarrow LR(h_2) = LR(g_3)LR(g_4) \tag{2.75}$$

leading to

$$LR(h_1) = 0.1 \quad LR(h_2) = 1$$

which can be written as likelihood vector as

$$L_2 = [0.1 \quad 1].$$

Finally, the likelihood in level-0, i.e., L_0, can be calculated as

$$LR(u_4) = LR(h_1)LR(h_2) \tag{2.76}$$

leading to

$$LR(u_4) = 0.1.$$

Using the decoding logic

$$u_i = \begin{cases} 0 \text{ if } L(u_i) \geq 1 \\ 1 \ otherwise \end{cases} \tag{2.77}$$

we can decide that the first decoded information bit is $u_4 = 1$. Together with the frozen bits the decoded data vector can be written as

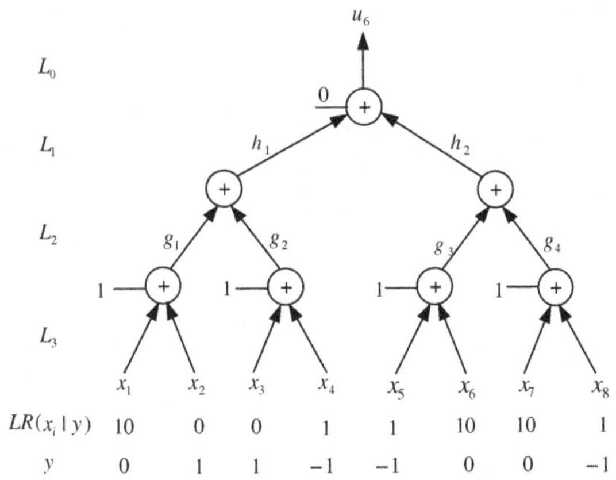

Fig. 2.49 Decoding tree for data bit u_6

$$\bar{u} = [u_1 = 0\, u_2 = 0\, u_3 = 0\, u_4 = 1].$$

Since the bit u_5 is a frozen bit, we can form the decoded data vector as

$$\bar{u} = [u_1 = 0\, u_2 = 0\, u_3 = 0\, u_4 = 1\, u_5 = 0]$$

and continue with the decoding of information bit u_6 as follows:

Decoding of u_6:

We first deliver the previously decoded information bits and frozen bits to the nodes. When the bits in the vector

$$\bar{u} = [u_1 = 0\, u_2 = 0\, u_3 = 0\, u_4 = 1\, u_5 = 0]$$

are delivered to the nodes as explained in Sect. 2.4, we obtain the node-bits in Fig. 2.49

The likelihoods in level-2, i.e., L_2, can be calculated as

$$\begin{aligned} LR(g_1) &= (LR(x_1))^{1-2\times 1} LR(x_2) \quad LR(g_2) = (LR(x_3))^{1-2\times 1} LR(x_4) \\ LR(g_3) &= (LR(x_5))^{1-2\times 1} LR(x_6) \quad LR(g_4) = (LR(x_7))^{1-2\times 1} LR(x_8) \end{aligned} \tag{2.78}$$

leading to

$$LR(g_1) = 0 \quad LR(g_2) = 10 \quad LR(g_3) = 10 \quad LR(g_4) = 0.1$$

which can be written as likelihood vector as

$$L_2 = [0 \quad 10 \quad 10 \quad 0.1].$$

The likelihoods in level-1, i.e., L_1, can be calculated as

$$LR(h_1) = \frac{1 + LR(g_1)LR(g_2)}{LR(g_1) + LR(g_2)} \quad LR(h_2) = \frac{1 + LR(g_3)LR(g_4)}{LR(g_3) + LR(g_4)} \tag{2.79}$$

leading to

$$LR(h_1) = 0.1 \quad LR(h_2) = 0.19$$

which can be written as likelihood vector as

$$L_2 = [0.1 \quad 0.19].$$

Finally, the likelihood in level-0, i.e., L_0, can be calculated as

$$LR(u_6) = (LR(h_1))^{1-2\times 0} LR(h_2) \rightarrow LR(u_6) = LR(h_1)LR(h_2) \tag{2.80}$$

leading to

$$LR(u_4) = 0.019.$$

Using the decoding logic

$$u_i = \begin{cases} 0 \text{ if } L(u_i) \geq 1 \\ 1 \; otherwise \end{cases} \tag{2.81}$$

we can decide that the decoded information bit is $u_6 = 1$. The decoded data vector can be written as

$$\bar{u} = [u_1 = 0 \, u_2 = 0 \, u_3 = 0 \, u_4 = 1 \, u_5 = 0 \, u_6 = 1].$$

Decoding of u_7:

We deliver the previously decoded information bits and frozen bits to the nodes. When the bits in the vector

$$\bar{u} = [u_1 = 0 \, u_2 = 0 \, u_3 = 0 \, u_4 = 1 \, u_5 = 0 \, u_6 = 1]$$

are delivered to the nodes as explained in Sect. 2.4, we obtain the node-bits in Fig. 2.50.

The likelihoods in level-2, i.e., L_2, can be calculated as

$$\begin{aligned} LR(g_1) &= (LR(x_1))^{1-2\times 1} LR(x_2) \quad LR(g_2) = (LR(x_3))^{1-2\times 1} LR(x_4) \\ LR(g_3) &= (LR(x_5))^{1-2\times 1} LR(x_6) \quad LR(g_4) = (LR(x_7))^{1-2\times 1} LR(x_8) \end{aligned} \tag{2.82}$$

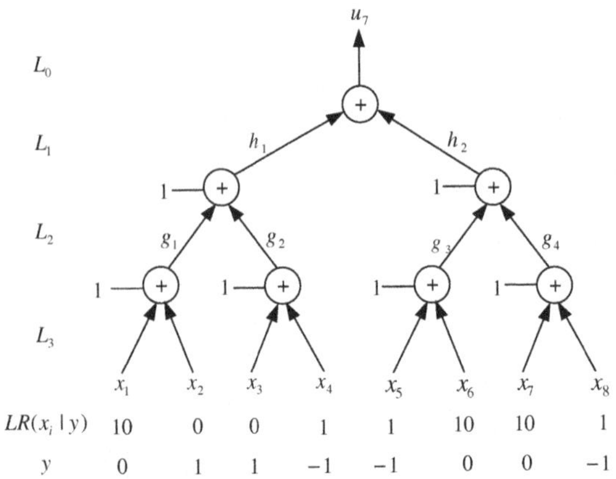

Fig. 2.50 Decoding tree for data bit u_7

leading to

$$LR(g_1) = 0 \quad LR(g_2) = 10 \quad LR(g_3) = 10 \quad LR(g_4) = 0.1$$

which can be written as likelihood vector as

$$L_2 = [0 \quad 10 \quad 10 \quad 0.1].$$

The likelihoods in level-1, i.e., L_1, can be calculated as

$$LR(h_1) = (LR(g_1))^{1-2\times 1} LR(g_2) \quad LR(h_2) = (LR(g_3))^{1-2\times 1} LR(g_4) \qquad (2.83)$$

leading to

$$LR(h_1) = 100 \quad LR(h_2) = 0.01$$

which can be written as likelihood vector as

$$L_2 = [100 \quad 0.01].$$

Finally, the likelihood in level-0, i.e., L_0, can be calculated as

$$LR(u_7) = \frac{1 + LR(h_1)LR(h_2)}{LR(h_1) + LR(h_2)}$$

leading to

$$LR(u_7) = 0.02.$$

Using the decoding logic

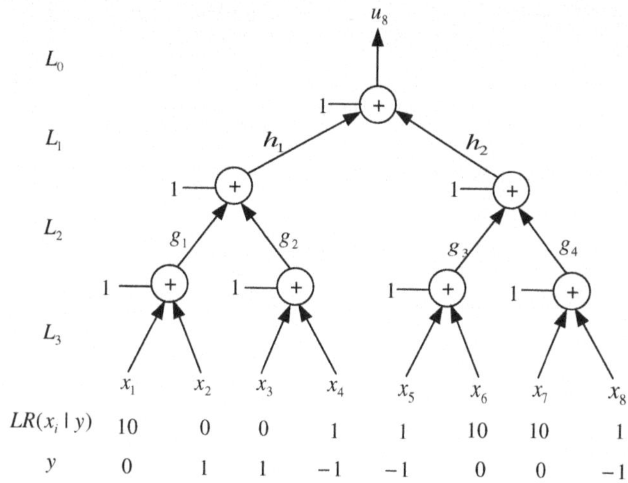

Fig. 2.51 Decoding tree for data bit u_8

$$u_i = \begin{cases} 0 \text{ if } L(u_i) \geq 1 \\ 1 \; otherwise \end{cases} \tag{2.84}$$

we can decide that the decoded information bit is $u_7 = 1$. The decoded data vector can be written as

$$\bar{u} = [u_1 = 0 \, u_2 = 0 \, u_3 = 0 \, u_4 = 1 \, u_5 = 0 \, u_6 = 1 \, u_7 = 1].$$

Decoding of u_8:

We deliver the previously decoded information bits and frozen bits to the nodes. When the bits in the vector

$$\bar{u} = [u_1 = 0 \, u_2 = 0 \, u_3 = 0 \, u_4 = 1 \, u_5 = 0 \, u_6 = 1 \, u_7 = 1].$$

are delivered to the nodes as explained in Sect. 2.4, we obtain the node-bits in Fig. 2.51.

The likelihoods in level-2, i.e., L_2, can be calculated similar to the decoding of bit u_7 as

$$L(g_1) = 0 \quad L(g_2) = 10 \quad L(g_3) = 10 \quad L(g_4) = 0.1$$

which can be written as likelihood vector as

$$L_2 = [0 \quad 10 \quad 10 \quad 0.1].$$

The likelihoods in level-1, i.e., L_1, can be calculated similar to the decoding of bit u_7 as

$$LR(h_1) = 100 \quad LR(h_2) = 0.01$$

which can be written as likelihood vector as

$$L_2 = [100 \quad 0.01].$$

Finally, the likelihood in level-0, i.e., L_0, can be calculated as

$$LR(u_8) = (LR(h_1))^{1-2\times 1} LR(h_1)$$

leading to

$$LR(u_8) = 0.0001.$$

Using the decoding logic

$$u_i = \begin{cases} 0 \text{ if } L(u_i) \geq 1 \\ 1 \ otherwise \end{cases} \tag{2.85}$$

we can decide that the decoded information bit is $u_8 = 1$. The decoded data vector can be written as

$$\bar{u} = [u_1 = 0\, u_2 = 0\, u_3 = 0\, u_4 = 1\, u_5 = 0\, u_6 = 1\, u_7 = 1\, u_8 = 1].$$

If we compare the decoded sequence to the original data vector including the frozen bits, we see that the decoded sequence is the same as the original information sequence.

Determination of Frozen Bit Locations for AWGN Channels

For the determination of split channel capacities for when AWGN channels are used for the transmission of code bits, we can still use (2.64) for the calculation of Bhattacharyya parameters of the split channels. However, in this case we should choose the initial value as

$$Z_1^1(W) = Z(W) = e^{-R\frac{E_b}{N_0}} \tag{2.86}$$

where R is the code rate. In addition to the method presented in (2.86), there are a number of techniques proposed in the literature for the determination of split channel capacities when AWGN channels are employed for the transmission.

Problems

(1) For the decoder structure in Fig. 2.8B, if $LR(x_1) = 10$, $LR(x_2) = 5$, $LR(x_3) = 10$ and $LR(x_4) = 0.1$, $LR(x_5) = 0.1$, $LR(x_6) = 0.1$, $LR(u_1)$, and $LR(u_2)$. Using $LR(u_1)$, and $LR(u_2)$ determine the bit values of u_1 and u_2 (Fig. P2.1).

(2) If $N = 128$, then how many levels are there in the decoder tree? How many nodes are available in each level?

(3) Let $N = 16$, and assume that the first 10 bits are decoded and the decoded bits are $u_1^{10} = [0\,0\,1\,1\,1\,0\,0\,1\,1\,0\,1]$. We want to decoder the 11th bit, i.e., u_{11}. Distribute the first decoded 10 bits to the tree nodes. Use generator matrices G_4 and G_2 for the bit distribution process (Fig. P2.2).

(4) For a binary erasure channel, if the erasure probability is $\alpha = 0.3$, calculate the channel capacities for $N = 8$.

(5) Encode the data vector $d = [1\,0\,1\,0\,1\,0\,1\,1]$ by a polar encoder and obtain the polar code-word.

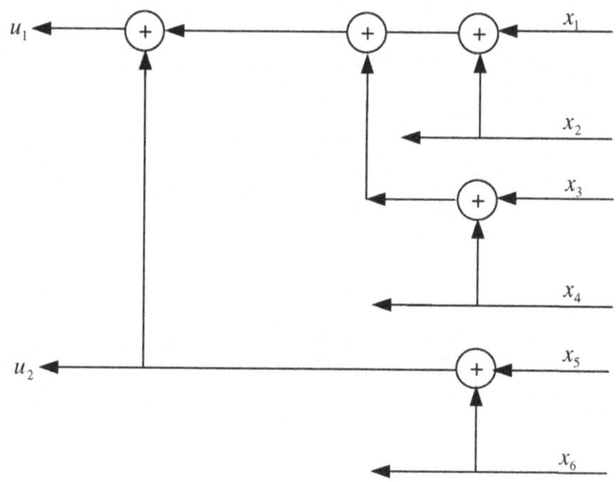

Fig. P2.1 Decoder segment

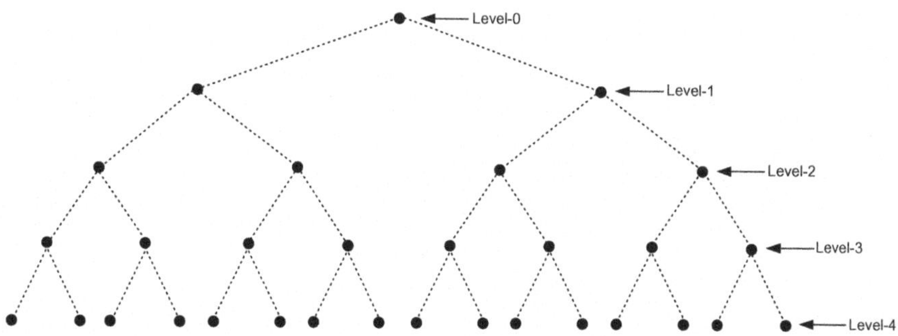

Fig. P2.2 Decoding tree and levels

(6) Consider the polar encoder depicted in Fig. 2.8. Assume that binary symmetric channels with transition probabilities $p(y = 1|x = 0) = 0.2$, $p(y = 0|x = 1) = 0.3$ is used for the transmission of code bits x_1^4. If the channel outputs are $y_1^4 = [0\,1\,1\,0]$, calculate $LR(x_i), i = 1, \ldots, 4$.

(7) Consider a polar encoder for $N = 8$, and assume that binary erasure channels with erasure probabilities $\alpha = 0.5$ are employed for the transmission of code bits and the code rate is $R = 0.5$. The channel outputs are given as

$$y_1^8 = [1 \quad 0 \quad -1 \quad 1 \quad 0 \quad -1 \quad 1 \quad 0].$$

Decode the code bits using the received signal vector y_1^8, and determine the data bits used in encoding operation.

Chapter 3
Polarization of Binary Erasure Channels

In this chapter, we will demonstrate the calculation of split channel capacities when polar codes are employed for binary erasure channels.

3.1 Polar Codes in Binary Erasure Channels

The binary erasure channel depicted in Fig. 3.1 where α is the erasure probability of the channel and it is defined as

$$\alpha = Prob(y = e|x = 0) \text{ or } \alpha = Prob(y = e|x = 1).$$

The capacity of the binary erasure channel is equal to

$$C = 1 - \alpha. \tag{3.1}$$

Any discrete channel is indicated is indicated by the letter W.

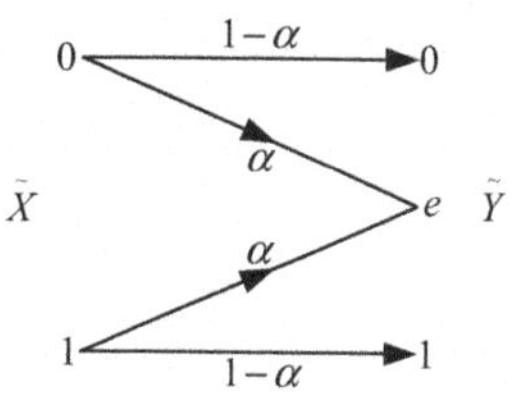

Fig. 3.1 Binary erasure channel with erasure probability α

O. Gazi, *Polar Codes*, Springer Topics in Signal Processing 15,
https://doi.org/10.1007/978-981-13-0737-9_3

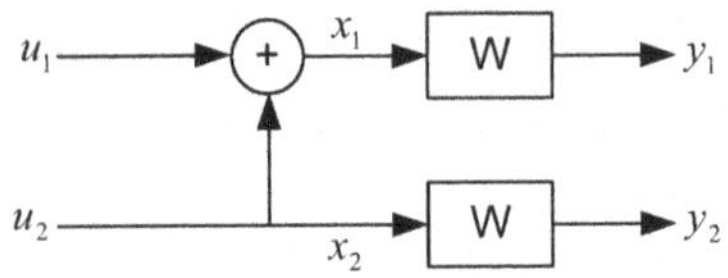

Fig. 3.2 Polar encoder structure for $N = 2$ together with two DMCs

3.1.1 Split Channels and Capacity of Split Channels When BECs are Employed

A discrete communication channel has a set of inputs and a set of outputs. Between every pair of output and input symbols, we have channel transition probabilities, and discrete channel is characterized by these set of transition probabilities. The data bits are coded and the code bits are transmitted through the discrete channel in a sequential manner.

Sequential transmission of code bits through the discrete channel W can also be considered as multiple use of the channel in a parallel manner as discussed in Chap. 1. Assume that there are N code bits transmitted through the identical channels in a parallel manner. The overall system between data bits and channel outputs can also be considered as a single channel having multiple inputs and multiple outputs. The decoding operation is a sequential process, and for the decoding of any data bit, we use all the available information at our hands. For the decoding of the first data bit, the available information is the all channel outputs. Then, using all the channel outputs we determine the value of the first data bit.

For the decoding of the second data bit, the available information that can be used is the channel output values, and the decision result of the first bit. Then, we can consider a channel whose input and outputs are the second data bit, and channel outputs together with the decision result of the first data bit.

In general, for the decoding of any data bit, we can define a channel whose input and outputs are the data bit to be determined and the channel outputs with the decision results of the previous bits. Such channels are called split channels.

In Fig. 3.2, a polar encoder structure for $N = 2$ together with two DMCs is depicted.

Assume that the discrete channels W shown in Fig. 3.2 are binary erasure channels, i.e., BECs. From Fig. 3.2, we can write

$$x_1 = u_1 \oplus u_2 \quad x_2 = u_2. \tag{3.2}$$

Using the structure of Fig. 3.2, we can define the split channels

$$\begin{aligned} W_2^1 &: u_1 \rightarrow y_1, y_2 \\ W_2^2 &: u_2 \rightarrow y_1, y_2, \hat{u}_1 \end{aligned} \tag{3.3}$$

where the second split channel uses the decision result of the first split channel at its output as available information together with the channel output.

Let's calculate the capacities of the split channels W_2^1 and W_2^2. The split channel W_2^1 and W_2^2 are also binary erasure channels. To determine the capacities of the split channels W_2^1 and W_2^2, we need to determine the erasure probabilities of the split channels. From (3.2), we can write that

$$
\begin{aligned}
u_1 &= x_1 \oplus x_2 \\
u_2 &= x_2
\end{aligned}
\tag{3.4}
$$

Now, let's calculate the erasure probabilities of the split channels.

Erasure probability of channel $W_2^1 : u_1 \rightarrow y_1, y_2$

Since $u_1 = x_1 \oplus x_2$, then we can state that u_1 is erased if any of the x_1 or x_2 is erased. In general, let $P_e(u_i)$ denote the erasure probability for the symbol $u_i = x_i \oplus x_{i+1} \oplus \cdots \oplus x_k$, and $P_e(u_i)$ can be calculated considering that $x_i, \ldots, x_j$ are erased and the symbols $x_{j+1}, \ldots, x_k$ are not erased. The bit u_1 is erased, if

$$
\begin{aligned}
&x_1 \, is \, erased, x_2 \, is \, not \, erased \\
&x_1 \, is \, not \, erased, x_2 \, is \, erased \\
&x_1 \, is \, erased, x_2 \, is \, erased
\end{aligned}
$$

and, we have

$$
\begin{aligned}
P_1 &= \{x_1 \, is \, erased, x_2 \, is \, not \, erased\} \rightarrow P_1 = \alpha \times (1-\alpha) \\
P_2 &= \{x_1 \, is \, not \, erased, x_2 \, is \, erased\} \rightarrow P_2 = (1-\alpha) \times \alpha \\
P_3 &= \{x_1 \, is \, erased, x_2 \, is \, erased\} \rightarrow P_3 = \alpha \times \alpha.
\end{aligned}
\tag{3.5}
$$

The erasure probability of u_1 equals to

$$
P_e(u_1) = P_1 + P_2 + P_2 \tag{3.6}
$$

leading to

$$
P_e(u_1) = \alpha \times (1-\alpha) + (1-\alpha) \times \alpha + \alpha \times \alpha \rightarrow P_e(u_1) = 2\alpha - \alpha^2
$$

Erasure probability of channel $W_2^2 : u_2 \rightarrow y_1, y_2, \hat{u}_1$

We can write that $u_2 = x_2$ or $u_2 = x_1 \oplus u_1$, then u_2 is erased if both x_1 and x_2 are erased. Then, we have

$$
\begin{aligned}
&P_1\{x_1 \, is \, erased, x_2 \, is \, erased\} \rightarrow P_1 = \alpha \times \alpha \\
&P_e(u_2) = P_1 \rightarrow P_e(u_2) = \alpha^2.
\end{aligned}
\tag{3.7}
$$

Let the erasure probability be equal to $\alpha = 0.4$, then we have

$$P_e(u_1) = 2 \times 0.4 - 0.4^2 \rightarrow P_e(u_1) = 0.64$$
$$\mathrm{P}_e\,(u_2) = 0.16.$$

Since we found the erasure probabilities of the channels, we can calculate the split channel capacities as

$$C(W_2^1) = 1 - P_e(u_1) \rightarrow C(W_2^1) = 1 - 0.64 \rightarrow C(W_2^1) = 0.36$$
$$C(W_2^2) = 1 - P_e(u_2) \rightarrow C(W_2^1) = 1 - 0.416 \rightarrow C(W_2^2) = 0.84$$

On the other hand, the capacity of the binary erasure channel having the erasure probability $\alpha = 0.4$ can be calculated as

$$C(W) = 1 - \alpha \rightarrow C(W) = 1 - 0.4 \rightarrow C(W) = 0.6.$$

And it is seen that

$$C(W_2^1) + C(W_2^2) = 2C(W) = 1.2.$$

Now let's calculate the split channel capacities for $N = 4$.

3.2 Capacity of Split Channels for $N=4$

To remind it, for $N = 4$, polar encoder structure is depicted in Fig. 3.3.

For the polar encoder structure of Fig. 3.3, we can define the split channels as

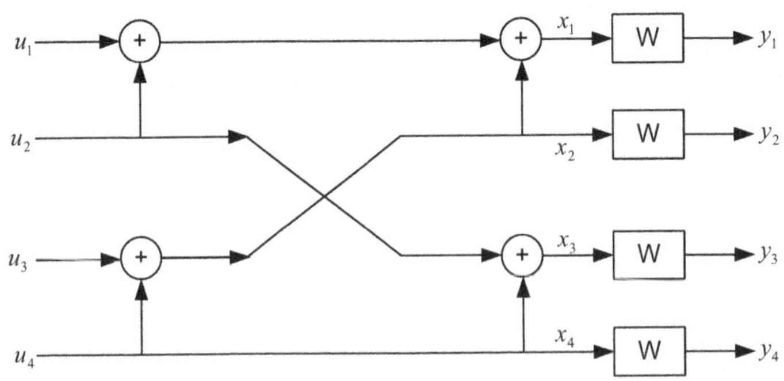

Fig. 3.3 Polar encoder structure for $N = 4$ together with four DMCs

$$
\begin{aligned}
W_4^1 &: u_1 \rightarrow y_1, y_2, y_3, y_4 \\
W_4^2 &: u_2 \rightarrow y_1, y_2, y_3, y_4, \hat{u}_1 \\
W_4^3 &: u_3 \rightarrow y_1, y_2, y_3, y_4, \hat{u}_1, \hat{u}_2 \\
W_4^4 &: u_4 \rightarrow y_1, y_2, y_3, y_4, \hat{u}_1, \hat{u}_2, \hat{u}_3
\end{aligned} \tag{3.8}
$$

which can be written in a simpler manner using the vectors, $y_1^4 = [y_1, y_2, y_3, y_4]$, and $u_i^k = [u_i, u_{i+1}, \cdots, u_k]$ as

$$
\begin{aligned}
W_4^1 &: u_1 \rightarrow y_1^4 \\
W_4^2 &: u_2 \rightarrow y_1^4, \hat{u}_1 \\
W_4^3 &: u_3 \rightarrow y_1^4, \hat{u}_1^2 \\
W_4^4 &: u_4 \rightarrow y_1^4, \hat{u}_1^3.
\end{aligned} \tag{3.9}
$$

For simplicity of notation, we can use y^4 for y_1^4 and $\hat{u}^k$ for $\hat{u}_1^k$. Then, we have

$$
\begin{aligned}
W_4^1 &: u_1 \rightarrow y^4 \\
W_4^2 &: u_2 \rightarrow y^4, \hat{u}_1 \\
W_4^3 &: u_3 \rightarrow y^4, \hat{u}^2 \\
W_4^4 &: u_4 \rightarrow y^4, \hat{u}^3.
\end{aligned} \tag{3.10}
$$

Using the structure of Fig. 3.3, we obtain

$$
\begin{aligned}
x_1 &= u_1 \oplus u_2 \oplus u_3 \oplus u_4 \\
x_2 &= u_3 \oplus u_4 \\
x_3 &= u_2 \oplus u_4 \\
x_4 &= u_4
\end{aligned} \tag{3.11}
$$

which can be written as

$$
\bar{x} = \bar{u}G \tag{3.12}
$$

where

$$
\bar{x} = [x_1 x_2 x_3 x_4] \quad \bar{u} = [u_1 u_2 u_3 u_4]
$$

and

$$
G = \begin{bmatrix} 1\,0\,0\,0 \\ 1\,0\,1\,0 \\ 1\,1\,0\,0 \\ 1\,1\,1\,1 \end{bmatrix}.
$$

The encoder matrix G of the polar code has the property

$$G = G^{-1} \tag{3.13}$$

which can be used in (3.12) to obtain $\bar{u}$ as

$$\bar{u} = \bar{x}G^{-1} \rightarrow \bar{u} = \bar{x}G \tag{3.14}$$

leading to

$$\begin{aligned} u_1 &= x_1 \oplus x_2 \oplus x_3 \oplus x_4 \\ u_2 &= x_3 \oplus x_4 \\ u_3 &= x_2 \oplus x_4 \\ u_4 &= x_4 \end{aligned} \tag{3.15}$$

from which, it is clear that

u_1 is lost if at least one symbol from the set x_1, x_2, x_3, x_4 is lost
u_2 is lost if at least one symbol from the set x_3, x_4 is lost
u_3 is lost if at least one symbol from the set x_2, x_4 is lost
u_4 is lost if x_4 is lost.

Then, for u_1 erasure or lose probability can be calculated as

$$\begin{aligned} P_e(u_1) = {} & P(x_1 \text{ is lost}) + P(x_2 \text{ is lost}) + P(x_3 \text{ is lost}) + P(x_4 \text{ is lost}) \\ & + P(x_1, x_2 \text{ are lost}) + P(x_1, x_3 \text{ are lost}) + P(x_1, x_4 \text{ are lost}) \\ & + P(x_2, x_3 \text{ are lost}) + P(x_2, x_4 \text{ are lost}) + P(x_3, x_4 \text{ are lost}) \\ & + P(x_1, x_2, x_3 \text{ are lost}) + P(x_1, x_3, x_4 \text{ are lost}) + P(x_2, x_3, x_4 \text{ are lost}) \\ & + P(x_1, x_2, x_3, x_4 \text{ are lost}) \end{aligned} \tag{3.16}$$

where

$$\begin{aligned} P(x_1 \text{ is lost}) = P(x_2 \text{ is lost}) = P(x_3 \text{ is lost}) = P(x_4 \text{ is lost}) = \alpha(1-\alpha)^3 \\ P(x_1, x_2 \text{ are lost}) = P(x_1, x_3 \text{ are lost}) = P(x_1, x_4 \text{ are lost}) = P(x_2, x_3 \text{ are lost}) \\ = P(x_2, x_4 \text{ are lost}) = P(x_2, x_3 \text{ are lost}) = \alpha^2(1-\alpha)^2 \\ P(x_1, x_2, x_3 \text{ are lost}) = P(x_1, x_3, x_4 \text{ are lost}) = P(x_2, x_3, x_4 \text{ are lost}) = \alpha^3(1-\alpha) \\ P(x_1, x_2, x_3, x_4 \text{ are lost}) = \alpha^4. \end{aligned} \tag{3.17}$$

Note: $P(x_1 \text{ is lost}) = P(\text{If only } x_1 \text{ is lost and } x_2, x_3, x_4 \text{ are not lost})$
Finally, using (3.17) in (3.16), $P_e(u_1)$ is found as

$$P_e(u_1) = 4\alpha(1-\alpha)^3 + 6\alpha^2(1-\alpha)^2 + 3\alpha^3(1-\alpha) + \alpha^4$$

which can be written as

$$P_e(u_1) = \sum_{i=1}^{4} \binom{4}{k} \alpha^k (1-\alpha)^{4-k}.$$

For u_2, erasure or lose probability can be calculated as

$$P_e(u_2) = P(x_3 \text{ is lost}) + P(x_4 \text{ is lost}) + P(x_3, x_4 \text{ are lost})$$

where

$$P(x_3 \text{ is lost}) = P(x_4 \text{ is lost}) = \alpha(1-\alpha)$$
$$P(x_3, x_4 \text{ are lost}) = \alpha^2.$$

Then, $P_e(u_2)$ is calculated as

$$P_e(u_2) = 2\alpha(1-\alpha) + \alpha^2.$$

For u_3, erasure or lose probability can be calculated as

$$P_e(u_3) = P(x_2 \text{ is lost}) + P(x_4 \text{ is lost}) + P(x_2, x_4 \text{ are lost})$$

where

$$P(x_2 \text{ is lost}) = P(x_4 \text{ is lost}) = \alpha(1-\alpha)$$
$$P(x_2, x_4 \text{ are lost}) = \alpha^2.$$

Then, $P_e(u_3)$ is calculated as

$$P_e(u_3) = 2\alpha(1-\alpha) + \alpha^2.$$

Lastly, for u_4, erasure or lose probability can be calculated as

$$P_e(u_3) = P(x_4 \text{ is lost}) = \alpha.$$

For $\alpha = 0.4$, numerical values of $P_e(u_1)$, $P_e(u_2)$, $P_e(u_2)$, and $P_e(u_4)$ can be found as

$$P_e(u_1) = 0.8320 \quad P_e(u_2) = 0.64 \quad P_e(u_2) = 0.64 \quad P_e(u_4) = 0.4.$$

Once, we have the channel erasure probabilities, we can calculate the capacities of the channels using

$$C = 1 - channel\ erasure\ probability$$

leading to

$$\begin{aligned}
&W_4^1 : C(W_4^1) = 1 - P_e(u_1) \rightarrow C(W_4^1) = 1 - 0.8320 \rightarrow C(W_4^1) = 0.1680 \\
&W_4^2 : C(W_4^2) = 1 - P_e(u_2) \rightarrow C(W_4^2) = 1 - 0.64 \rightarrow C(W_4^2) = 0.36 \\
&W_4^3 : C(W_4^3) = 1 - P_e(u_3) \rightarrow C(W_4^3) = 1 - 0.64 \rightarrow C(W_4^3) = 0.36 \\
&W_4^4 : C(W_4^4) = 1 - P_e(u_4) \rightarrow C(W_4^4) = 1 - 0.4 \rightarrow C(W_4^4) = 0.6.
\end{aligned} \tag{3.18}$$

If we sort the split channel probabilities in ascending order, we see that

$$C(W_4^1) < C(W_4^2) \leq C(W_4^3) < C(W_4^4).$$

As $N \rightarrow \infty$, split channel probabilities converges to either to zero or to one, i.e., as $N \rightarrow \infty$, we have

$$C(W_N^i) \approx 0 \,\text{or}\, C(W_N^i) \approx 1. \tag{3.19}$$

The proof of the above claim will be provided in Chap. 5.

3.3 Bhattacharyya Parameter

For a binary discrete memoryless channel W, the Bhattacharyya parameter is defined as

$$Z(W) = \sum_y \sqrt{Prob(y|x=0)Prob(y|x=1)} \tag{3.20}$$

which takes values in the range [0 1]. Bhattacharyya parameter and mutual information are inversely related to each other. That is, if Bhattacharyya parameter takes its maximum value, the mutual information takes its minimum value and vice versa. The Bhattacharyya parameter can also be considered as an upper bound for the maximum probability of bit error.

Example 3.1 Calculate the Bhattacharyya parameter for the binary erasure channel.

Solution 3.1 When the summation term in the Bhattacharyya parameter

$$Z(W) = \sum_y \sqrt{Prob(y|x=0)Prob(y|x=1)}$$

is expanded for BEC, we get

$$Z(W) = \sqrt{Prob(y=0|x=0)\underbrace{Prob(y=0|x=1)}_{=0}}$$

$$+\sqrt{\underbrace{Prob(y=1|x=0)}_{=0}\,Prob(y=1|x=1)}$$
$$+\sqrt{\underbrace{Prob(y=e|x=0)}_{\alpha}\,\underbrace{Prob(y=e|x=1)}_{\alpha}}$$
$$=\alpha$$

Thus,

$$Z(W)=\alpha.$$

The capacity of BEC is $C(W)=1-\alpha$, then we have

$$C(W)+Z(W)=1.$$

Example 3.2 Calculate the Bhattacharyya parameter for the binary symmetric channel.

Solution 3.2 For the binary symmetric channel, we have

$$Prob(y=0|x=0)=Prob(y=1|x=1)=1-\alpha$$
$$Prob(y=1|x=0)=Prob(y=0|x=1)=\alpha.$$

When the summation term in the Bhattacharyya parameter

$$Z(W)=\sum_{y}\sqrt{Prob(y|x=0)Prob(y|x=1)}$$

is expanded for BSC, we get

$$Z(W)=2\sqrt{\alpha(1-\alpha)}.$$

The capacity of the binary symmetric channel is

$$C(W)=1+\alpha\log_2\alpha+(1-\alpha)\log_2(1-\alpha)\rightarrow C(W)=1-h_b(\alpha)$$

where

$$h_b(\alpha)=-\alpha\log_2\alpha-(1-\alpha)\log_2(1-\alpha).$$

The graphs of $Z(W)$ and $C(W)$ for the binary symmetric channel are depicted in Fig. 3.4.

Exercise *Calculate the Bhattacharyya parameter for the channel shown in Fig. 3.5 where $x_1=y_1=0$ and $x_2=y_2=1$.*

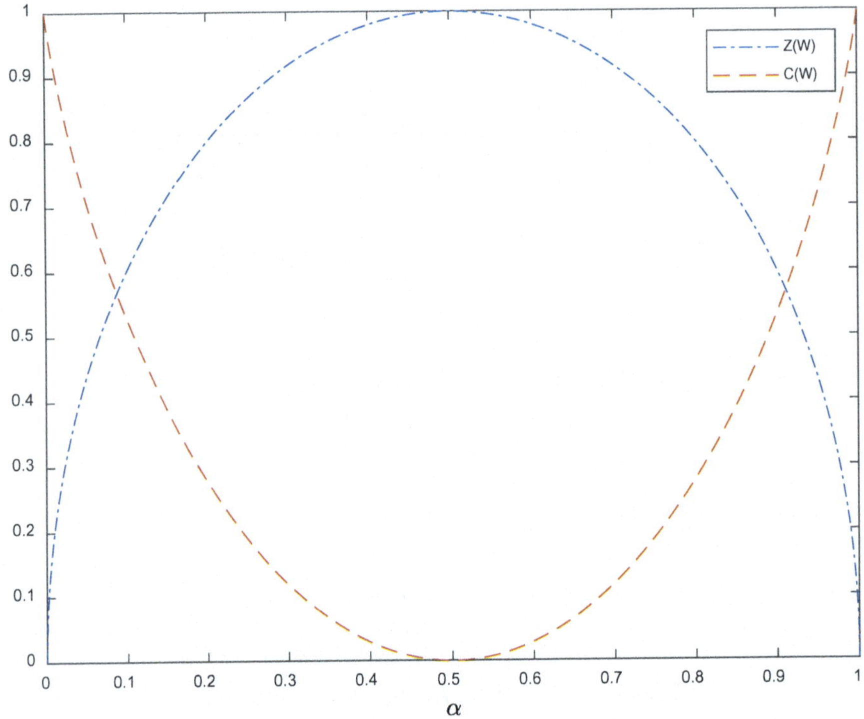

Fig. 3.4 The graphs of $Z(W)$ and $C(W)$ for the binary symmetric channel

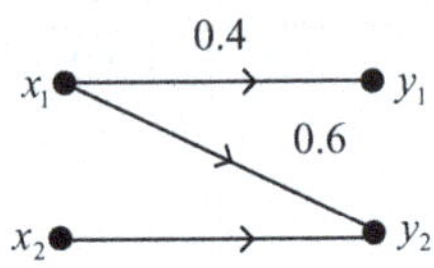

Fig. 3.5 A discrete memoryless channel

3.4 Capacity Bound Using Bhattacharyya Parameter

The channel capacity satisfies the bound

$$\log_2 \frac{2}{1 + Z(W)} \leq C(W) \leq \sqrt{1 - Z(W)^2}. \tag{3.21}$$

The graph of (3.21) for binary symmetric channel is depicted in Fig. 3.6.

Example 3.3 Verify the inequality (3.21) for binary erasure channel.

Solution 3.3 For binary erasure channel, we have

$$C(W) = 1 - \alpha \quad \text{and} \quad Z(W) = \alpha.$$

For binary erasure channel, the inequality (3.21) takes the form

$$\log_2 \frac{2}{1+\alpha} \leq 1 - \alpha \leq \sqrt{1 - \alpha^2}.$$

Exercise *Calculate the mutual information of the binary symmetric channel and verify the inequality (3.21).*

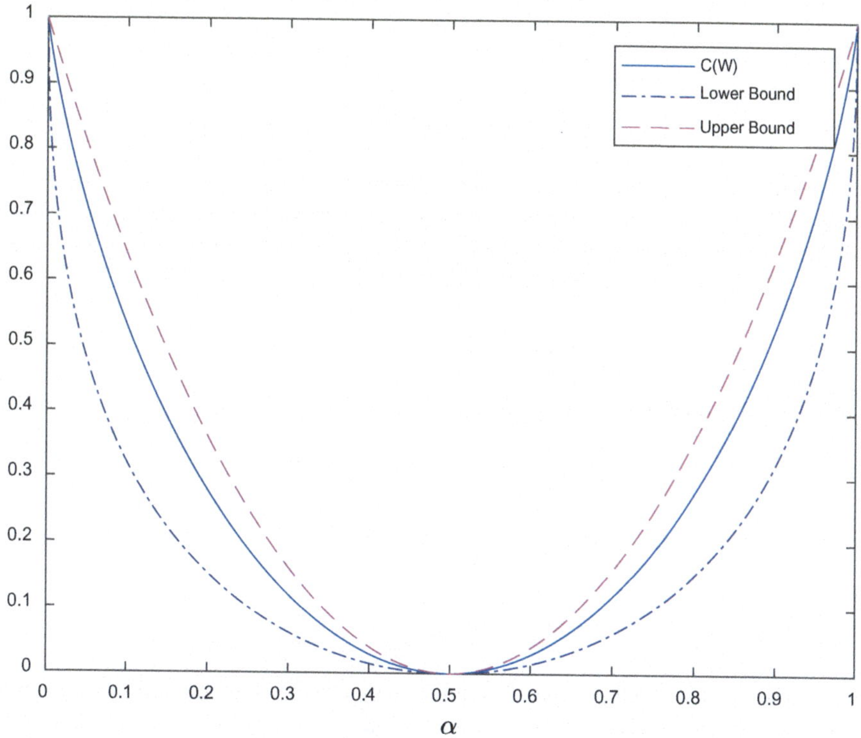

Fig. 3.6 The graph of (3.21) for binary symmetric channel

3.5 Inequalities for the Mutual Information and Bhattacharyya Parameters of the Split Channels

Mutual Information and Bhattacharyya Parameter for Split Channels:

The kernel unit of the polar encoder system is repeated below for reminder.

For the system in Fig. 3.7, the split channels can be written as

$$W_2^1 : u_1 \rightarrow y_1, y_2$$
$$W_2^2 : u_2 \rightarrow y_1, y_2, \hat{u}_1 \tag{3.22}$$

and we previously showed that for BEC, we have

$$C(W_2^1) + C(W_2^2) = 2C(W)$$

and

$$C(W_2^1) < C(W_2^2).$$

This idea can be generalized for split channels in general. Channel splitting does not alter the total capacity of the communication channels. Splitting causes the channel capacities to converge either to 0 or to 1. In general, for the ith split channel, the mutual information, or channel capacity satisfy

$$C(W_{2N}^{2i-1}) + C(W_{2N}^{2i}) = 2C(W_N^i) \tag{3.23}$$

and reliability improves in the sense that

$$Z(W_{2N}^{2i-1}) + Z(W_{2N}^{2i}) \leq 2Z(W_N^i) \tag{3.24}$$

where equality occurs for binary erasure channels.

For the capacity expressions in (3.23), we have the inequality

$$C(W_{2N}^{2i-1}) \leq C(W_N^i) \leq C(W_{2N}^{2i}) \tag{3.25}$$

which implies that channel capacities move away from each other, and show a convergence behavior for very large N values. Since the capacity and Bhattacharyya parameter are oppositely related to each other, (3.25) implies that

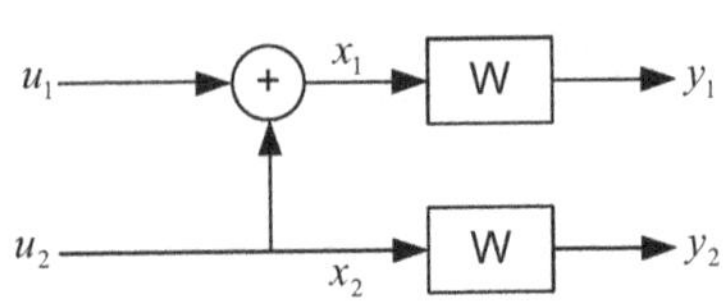

Fig. 3.7 The kernel unit of the polar encoder with two DMCs

$$Z(W_{2N}^{2i-1}) \geq Z(W_N^i) \geq Z(W_{2N}^{2i}). \tag{3.26}$$

In addition, for the Bhattacharyya parameter, we have the inequality

$$Z(W_{2N}^{2i-1}) \leq 2Z(W_N^i) - (Z(W_N^i))^2 \tag{3.27}$$

and equality

$$Z(W_{2N}^{2i}) = (Z(W_N^i))^2 \tag{3.28}$$

and for binary erasure channels equality occurs in (3.27). The sum of the channel capacities of the split channels satisfy

$$\sum_i C(W_N^i) = NC(W) \tag{3.29}$$

and, in a similar manner the sum of the Bhattacharyya parameters the split channels satisfy

$$\sum_i Z(W_N^i) \leq NZ(W). \tag{3.30}$$

3.6 Split Binary Erasure Channels

In Sect. 3.1.1, we showed that for $N = 2$, the erasure probability of the split channel W_2^1 equals to $P_e(u_1) = 2\alpha - \alpha^2$ where α is the erasure probability of the binary erasure channel W. In addition, we also showed that the erasure probability of the split channel W_2^2 equals to $P(u_2) = \alpha^2$.

In fact, the calculated erasure probabilities are nothing but the Bhattacharyya parameters of the corresponding channels. Note that Bhattacharyya parameter of a split channel can be considered as an upper bound for the maximum bit error probability for the corresponding channel.

If we denote the Bhattacharyya parameter of W by $Z(W) = \alpha$, and denote the channel W by W_1^1, then the Bhattacharyya parameter of W_2^1, i.e., $Z(W_2^1) = 2\alpha - \alpha^2$ and the Bhattacharyya parameter of $Z(W_2^2) = \alpha^2$ can be written in terms of $Z(W_1^1)$ as in

$$Z(W_2^1) = 2Z(W_1^1) - (Z(W_1^1))^2$$
$$Z(W_2^2) = (Z(W_1^1))^2$$

which can be generalized for the split binary erasure channels for any N value as in

$$\begin{aligned} Z(W_N^{2i-1}) &= 2Z(W_{N/2}^i) - (Z(W_{N/2}^i))^2 \\ Z(W_N^{2i}) &= (Z(W_{N/2}^i))^2 \end{aligned} \tag{3.31}$$

where $Z(W_1^1) = \alpha$. Since capacity and Bhattacharyya parameter are oppositely related to each other, i.e., $C(W_N^i) = 1 - Z(W_N^i)$, then (3.31) implies that

$$\begin{aligned} C(W_N^{2i}) &= 2C(W_{N/2}^i) - (C(W_{N/2}^i))^2 \\ C(W_N^{2i-1}) &= (C(W_{N/2}^i))^2 \end{aligned} \tag{3.32}$$

where $C(W_1^1) = 1 - \alpha$.

The recursive equations given in (3.32) can be calculated using the matlab program in Prog. 3.1.

```
clc; clear all;
indx=1:10;
N=2.^(indx);
C(1,1)=0.5;
for indx1=1:1:10
   N0=N(1,indx1);
   for j=1:N0/2
    C(N0,2*j-1)=C(N0/2,j)^2;
    C(N0,2*j)=2*C(N0/2,j)-C(N0/2,j)^2;
   end
end
N0=128;
plot(1:N0,C(N0,1:N0),'k*');
axis([1 N0, 0 1]);
xlabel('i');
ylabel('C_{128}^i');
hold on;

N0=1024;
plot(1:N0,C(N0,1:N0),'k*');
axis([1 N0, 0 1]);
xlabel('i');
ylabel('C_{1024}^i');
hold on;
```

Prog. 3.1 Calculation of the recursive equations in (3.32).

The graph of (3.32) for the erasure probability $\alpha = 0.5$ is depicted in Fig. 3.8 for $N = 128$, and $N = 1024$.

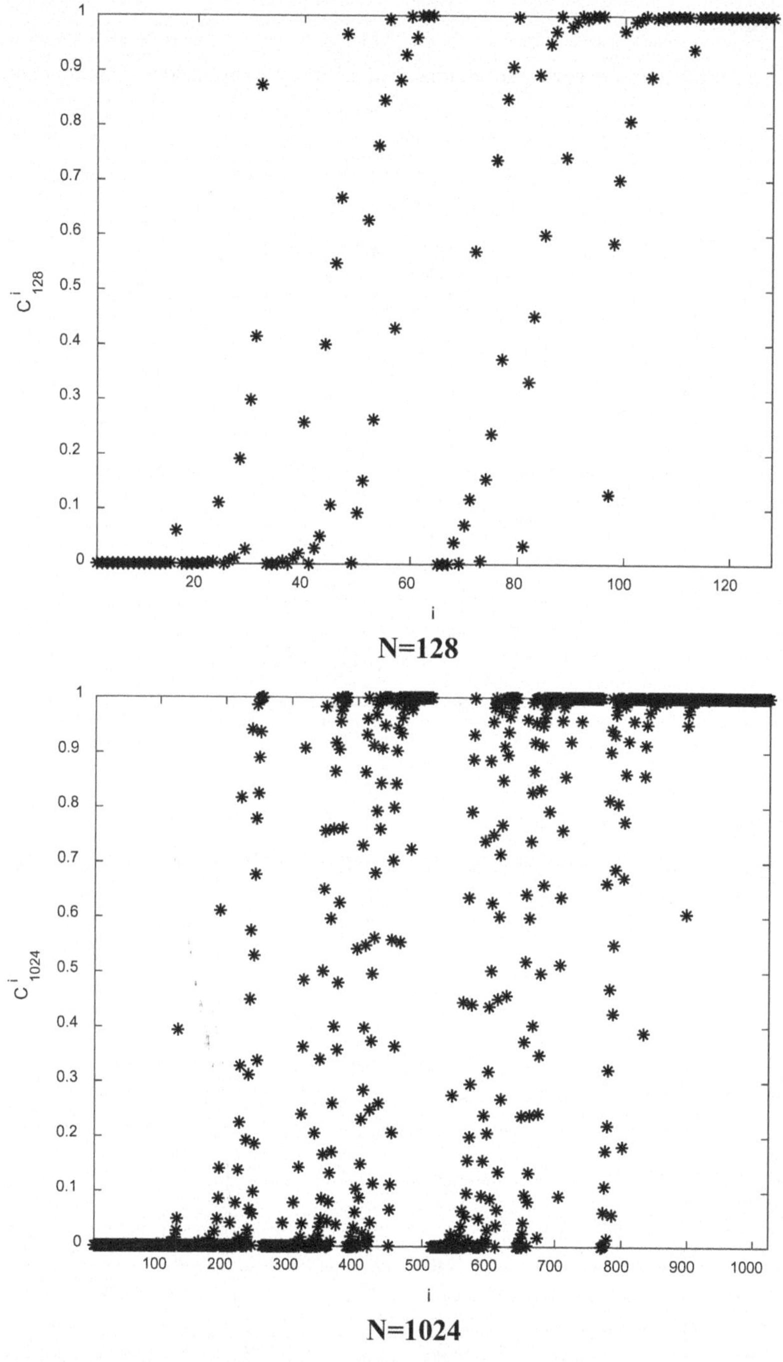

Fig. 3.8 Split channel capacities for $N = 128$ and $N = 1024$ when BECs with $\alpha = 0.5$ are employed for transmission

Problems

(1) Find the Bhattacharyya parameters of the binary memoryless channels shown in Fig. P3.1

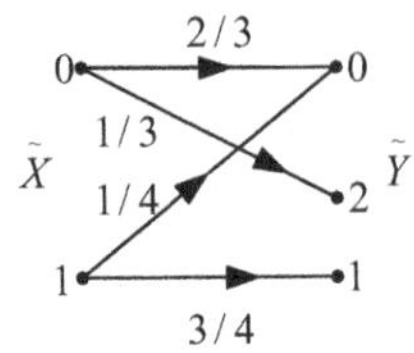

Fig. P3.1 Discrete memoryless channel for P1

(2) Calculate the erasure probability of the concatenated channel in Fig. P3.2, and calculate the capacity of the channel.

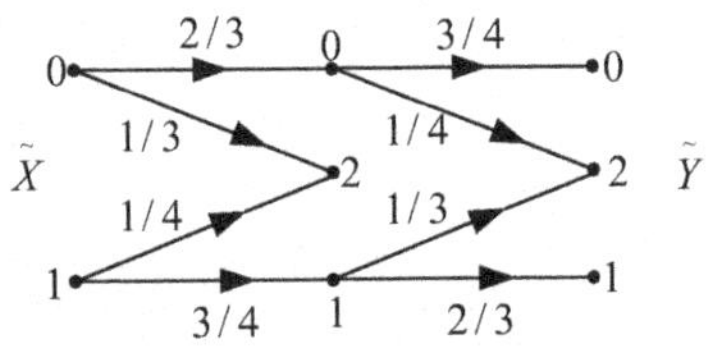

Fig. P3.2 Concatenated BECs for P2

(3) For the binary memoryless channel shown in Fig. P3.3, write and upper bound for the probability of bit error.

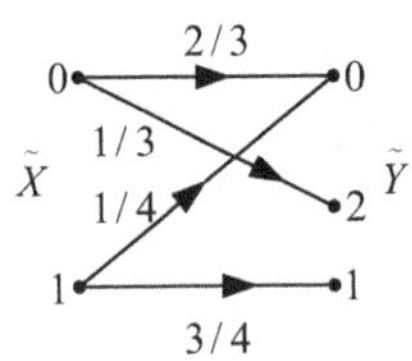

Fig. P3.3 Discrete memoryless channel for P3

(4) Indicate the split channels for the polar encoder for $N = 8$.

(5) Calculate the probability of error for the split channels of the polar encoders for $N = 2$ and for $N = 4$.

(6) For $N = 4$, assume that binary erasure channels with erasure probability $\alpha = 0.5$ are used for the transmission of polar encoded bits. Calculate the Bhattacharyya parameters of the split channel W_4^1, W_4^2, W_4^3, and W_4^4 using the definition of Bhattacharyya parameter in (3.20).

Chapter 4
Mathematical Modelling of Polar Codes, Channel Combining and Splitting

In Chap. 2, we explained the decoding of polar codes in a clear and trivial manner. However, in most of the engineering sciences, it is critical to mathematically formulate the theories or techniques invented. In this chapter, we will explain the mathematical formulization of successive cancellation decoding of polar codes. We first derive the necessary formulas for the successive cancellation decoding of polar codes in a non-recursive manner, then derive the formulas for the recursive successive cancellation decoding of polar codes which is a direct consequence of recursive channel construction.

4.1 Channel Combining

In this section, we will explain the idea of channel combining, and inspect the combined system in details.

A discrete channel has a number of input symbols and a number of output symbols, and the channel is identified by the transition probabilities between channel outputs and inputs. A discrete channel with a single input and a single output is shown in Fig. 4.1.

Assume that a number of bits are sequentially transmitted through the channel as shown in Fig. 4.2.

The transmission of N bits through the channel corresponds to the N use of the channel, i.e., each transmission can be considered as a separate use of the same

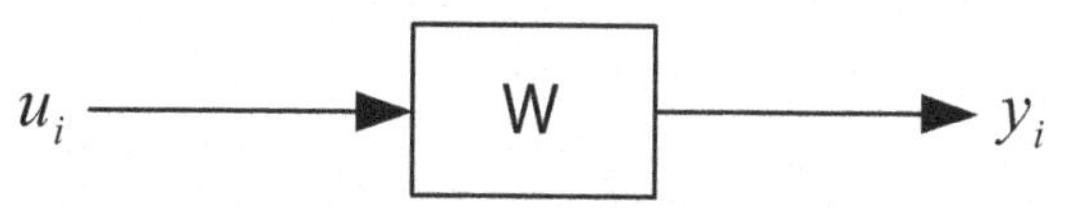

Fig. 4.1 A discrete channel with a single input and a single output

O. Gazi, *Polar Codes*, Springer Topics in Signal Processing 15,
https://doi.org/10.1007/978-981-13-0737-9_4

channel. Thus, an equivalent transmission scheme of Fig. 4.2 can be drawn as in Fig. 4.3.

From Fig. 4.3, we see that there are N identical discrete memoryless channel. Now, let's encode the data bits $\bar{u} = \left[u_1, u_2 \ldots, u_{N-1}, u_N\right]$ through the use of

$$\bar{x} = \bar{u}G \tag{4.1}$$

where G is the polar encoder matrix, and send the code bits $\bar{x} = [x_1, \ldots, x_N]$ through the identical channels as shown in Fig. 4.4.

We can consider the transmission scheme in Fig. 4.4 as a transmission through a single channel such that the channel inputs are $\bar{u} = \left[u_1, u_2 \ldots, u_{N-1}, u_N\right]$, and the channel outputs are $y_1^N = \left[y_1\ y_2 \ldots y_{N-1}\ y_N\right]$, then we call this channel "combined channel" and indicate it by W_N meaning that N separate discrete channels, i.e., W, are considered together. The channel combining and its black-box representation are illustrated in Fig. 4.4.

4.1.1 Probability Background Information

In this subsection, we will provide some probability background information. Let $\tilde{X}$ and $\tilde{Y}$ be discrete random variables with marginal and joint probability mass functions $p_x(x)$, $p_y(y)$, and $p_{xy}(x, y)$ respectively, and let $g(\cdot)$ be a function such that $\tilde{Z} = g(\tilde{X})$. If $z = g(x)$, then we have

$$p_x(x) = p_z(z) \text{ i.e. } Prob(\tilde{X} = x) = Prob(\tilde{Z} = z) \tag{4.2}$$

and

$$p_x(x) = \sum_y p_x(x, y) \quad p_y(y) = \sum_x p_x(x, y). \tag{4.3}$$

In addition, we have

$$p_{xy}(x, y) = p_{zy}(z, y) \quad \text{i.e. } p_{xy}(x, y) = p_{zy}(g(x), y) \tag{4.4}$$

which can be written considering more random variables as

$$p_{xy}(x_1, x_2, y_1, y_2) = p_{zy}(z_1, z_2, y_1, y_2) \tag{4.5}$$

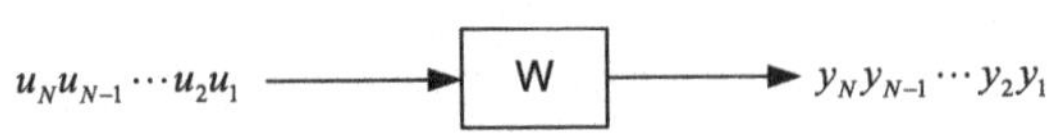

Fig. 4.2 Sequential transmission of bits through DMC

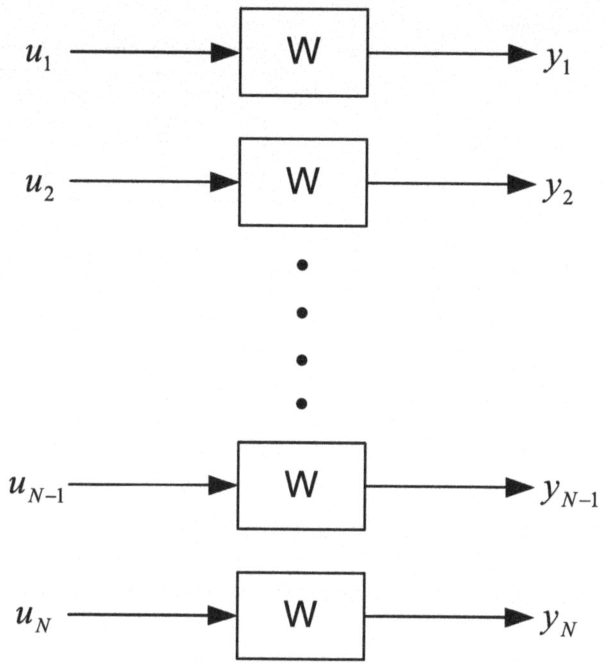

Fig. 4.3 Parallel DMCs

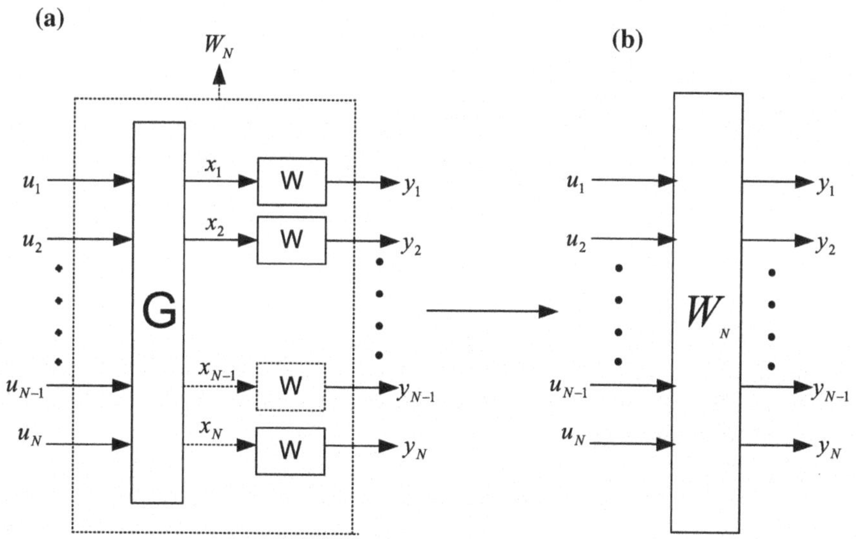

Fig. 4.4 Combined channel W_N

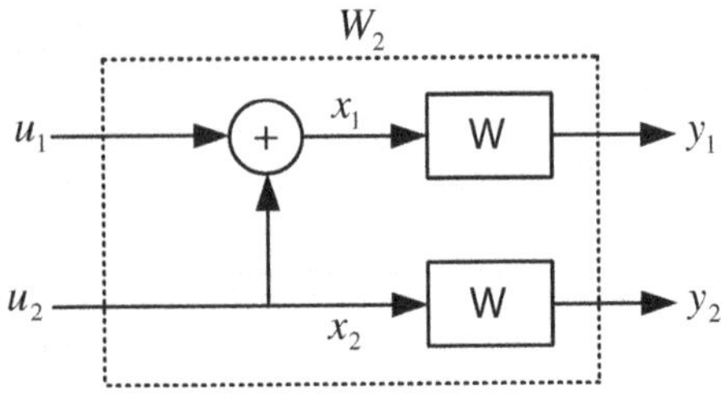

Fig. 4.5 Combined channel W_2

where

$$z_1 = g_1(x_1, x_2) \quad z_2 = g_2(x_1, x_2) \tag{4.6}$$

i.e.,

$$p_{xy}(x_1, x_2, y_1, y_2) = p_{zy}(g_1(x_1, x_2), g_2(x_1, x_2), y_1, y_2) \tag{4.7}$$

If $\tilde{X}$ and $\tilde{Y}$ are independent discrete random variables, then we have

$$p_{xy}(x, y) = p_x(x)p_y(y). \tag{4.8}$$

Example 4.1 For the W_2 channel shown in Fig. 4.5, show that

$$p(u_1, u_2, y_1, y_2) = p(x_1, x_2, y_1, y_2) \tag{4.9}$$

Solution 4.1 From Fig. 4.5, it is obvious that

$$x_1 = g_1(u_1, u_2) \quad x_2 = g_2(u_1, u_2)$$

then we have,

$$p(u_1, u_2, y_1, y_2) = p(x_1, x_2, y_1, y_2).$$

If the random variables $\tilde{U}_1$ and $\tilde{U}_2$ generates u_1 and u_2, then it is obvious that the random variables $\tilde{X}_1$ and $\tilde{X}_2$ generates $u_1 \oplus u_2$ and u_2, since $\tilde{X}_1 = \tilde{U}_1 \oplus \tilde{U}_2$ and $\tilde{X}_2 = \tilde{U}_2$. This means that

$$p(u_1, u_2) = p(x_1, x_2)$$

i.e.,

$$Prob(\tilde{U}_1 = u_1, \tilde{U}_2 = u_2) = Prob(\tilde{X}_1 = x_1, \tilde{X}_2 = x_2)$$

Table 4.1 Data and code bits

u_1	u_2	$p(u_1)$	$p(u_2)$	$x_1 = u_1 \oplus u_2$	$x_2 = u_2$
0	0	1/2	1/2	0	0
0	1	1/2	1/2	1	1
1	0	1/2	1/2	1	0
1	1	1/2	1/2	0	1

leading to

$$p(u_1, u_2, y_1, y_2) = p(x_1, x_2, y_1, y_2).$$

Example 4.2 For the encoder of Fig. 4.5, show that

$$p(u_1, u_2) = p(u_1)p(u_2) \quad p(x_1, x_2) = p(x_1)p(x_2) \tag{4.10}$$

Solution 4.2 From Fig. 4.5, It is obvious that u_1 and u_2 are independent of each other, then we have

$$p(u_1, u_2) = p(u_1)p(u_2).$$

In Table 4.1, the all possible values of u_1, u_2, x_1, x_2 together with $p(u_1)$, $p(u_2)$ are depicted.

From Table 4.1, we can write that

$$p(x_1 = 0) = p(u_1 = 0)p(u_2 = 0) + p(u_1 = 1)p(u_2 = 1) \rightarrow p(x_1 = 0) = \frac{1}{2}$$

$$p(x_1 = 1) = p(u_1 = 0)p(u_2 = 1) + p(u_1 = 1)p(u_2 = 0) \rightarrow p(x_1 = 1) = \frac{1}{2}$$

and

$$p(x_2 = 0) = p(u_2 = 0) \rightarrow p(x_2 = 0) = \frac{1}{2}$$

$$p(x_2 = 1) = p(u_2 = 1) \rightarrow p(x_2 = 1) = \frac{1}{2}.$$

In addition, From Table 4.1 we see that

$$p(x_1 = 0, x_2 = 0) = p(u_1 = 0)p(u_2 = 0)$$

and $p(x_1 = 0, x_2 = 0)$ satisfies

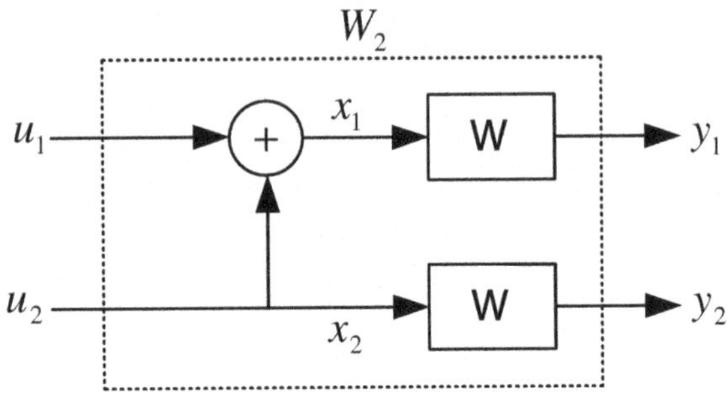

Fig. 4.6 W_2 channel for Example 4.3

$$p(x_1 = 0, x_2 = 0) = p(x_1 = 0)p(x_2 = 0).$$

Similarly, we can show that

$$p(x_1 = 0, x_2 = 1) = p(x_1 = 0)p(x_2 = 1)$$
$$p(x_1 = 1, x_2 = 0) = p(x_1 = 1)p(x_2 = 0)$$
$$p(x_1 = 1, x_2 = 1) = p(x_1 = 1)p(x_2 = 1).$$

Thus, we see that the random variables $\tilde{X}_1$ and $\tilde{X}_2$ are independent of each other, and we can write that

$$p(x_1, x_2) = p(x_1)p(x_2).$$

In general, we can write that

$$p(u_1, u_2, \ldots, u_N) = p(u_1)p(u_2)\ldots p(u_N)$$

$$p(x_1, x_2, \ldots, x_N) = p(x_1)p(x_2)\ldots p(x_N).$$

Example 4.3 For the W_2 channel shown in Fig. 4.6, show that

$$Prob(y_1, y_2|u_1, u_2) = Prob(y_1|u_1 \oplus u_2)Prob(y_2|u_2) \tag{4.11}$$

which can be written as

$$W_2(y_1^2|u_1^2) = W(y_1|u_1 \oplus u_2)W(y_2|u_2) \tag{4.12}$$

where we can also use W_1 for the place of W.

Solution 4.3 From Example 4.1, we obtained that

$$p(u_1, u_2, y_1, y_2) = p(x_1, x_2, y_1, y_2)$$

which can be written as

$$p(y_1, y_2|u_1, u_2)p(u_1, u_2) = p(y_1, y_2|x_1, x_2)p(x_1, x_2)$$

where

$$p(u_1, u_2) = p(u_1)p(u_2) \rightarrow p(u_1, u_2) = \frac{1}{2} \times \frac{1}{2} \rightarrow p(u_1, u_2) = \frac{1}{4}$$

$$p(x_1, x_2) = p(x_1)p(x_2) \rightarrow p(x_1, x_2) = \frac{1}{2} \times \frac{1}{2} \rightarrow p(x_1, x_2) = \frac{1}{4}$$

then we have

$$p(y_1, y_2|u_1, u_2) = p(y_1, y_2|x_1, x_2) \tag{4.13}$$

Since x_1 and x_2 are independent of each other, then (4.13) can be written as

$$p(y_1, y_2|u_1, u_2) = p(y_1|x_1)p(y_2|x_2)$$

in which substituting

$$x_1 = u_1 \oplus u_2 \quad x_2 = u_2$$

we get

$$p(y_1, y_2|u_1, u_2) = p(y_1|u_1 \oplus u_2)p(y_2|u_2)$$

which can be written as

$$W_2(y_1^2|x_1^2) = W(y_1|u_1 \oplus u_2)W(y_2|u_2).$$

4.1.2 Recursive Construction of Polar Encoder Structures

The recursive logic used for the construction of polar encoder structures is depicted in Fig. 4.7 for reminder. We will use this structure while deriving the formulas for the successive cancellation decoding of polar codes. First, we will derive some formulas for the combined channel, next, we will consider the decoding operation using the split channels referring to the recursive encoder structure depicted in Fig. 4.7.

Before studying the channel probabilities, let's introduce the notation used in formulation. The row vector $(x_1, x_2, \ldots, x_N)$ is indicated by x_1^N, and $x_i^j, j > i$ shows the sub-vector $\left(x_i, x_{i+1}, \cdots, x_j\right)$, and $x_{1,o}^K$ indicates the sub-vector with odd indices, similarly, $x_{1,e}^K$ indicates the sub-vector with even indices.

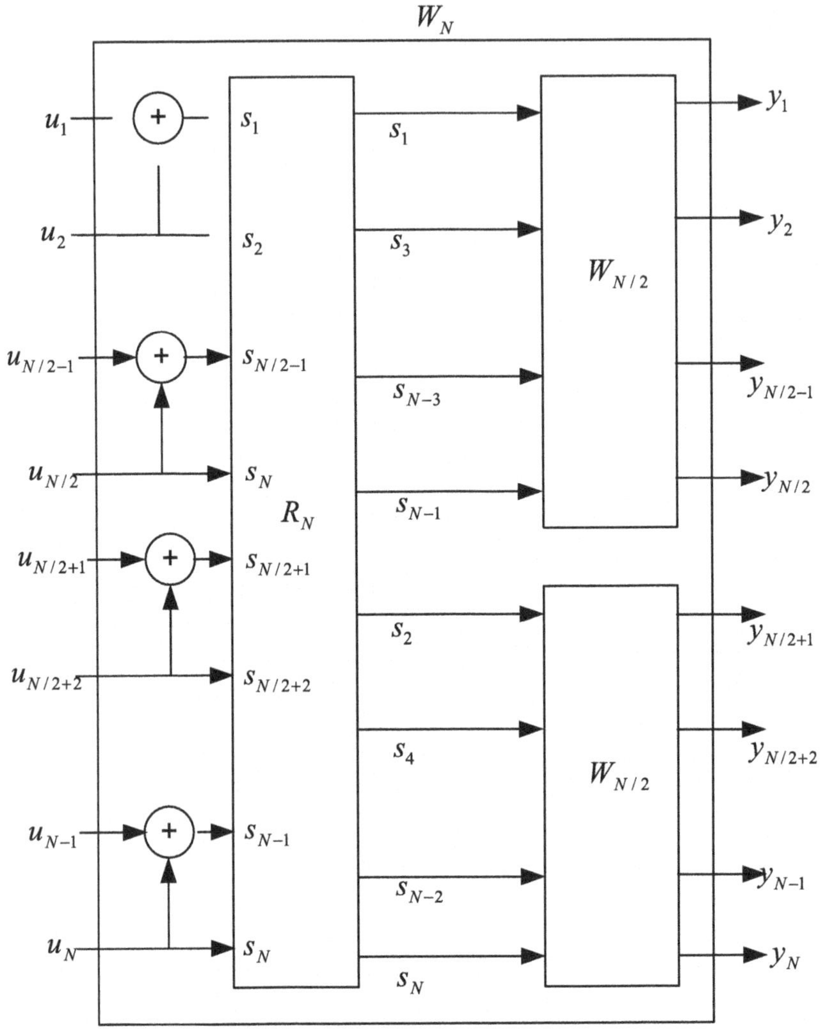

Fig. 4.7 Recursive construction of polar encoders

Example 4.4 If $x_1^9 = [1\ 2\ 3\ 4\ 5\ 6\ 7\ 8\ 9]$, then we have

$$x_3^8 = [3\ 4\ 5\ 6\ 7\ 8] \quad x_{3,e}^6 = [4\ 6] \quad x_{1,o}^7 = [1\ 3\ 5\ 7].$$

Example 4.5 Write the explicit form of

$$u_{1,o}^{2N} \oplus u_{1,e}^{2N}. \tag{4.14}$$

Solution 4.5 If the data vector is written as

$$u_1^{2N} = [u_1\ u_2 \ldots u_{2N}]$$

then we have

$$u_{1,o}^{2N} = \left[u_1\ u\ u_5 \ldots u_{2N-1}\right]\ \ u_{1,e}^{2N} = [u_2\ u_4\ u_6 \ldots u_{2N}]$$

and

$$u_{1,o}^{2N} \oplus u_{1,e}^{2N} = \left[u_1 \oplus u_2\ u_3 \oplus u_4\ u_5 \oplus u_6 \ldots u_{2N-1} \oplus u_{2N}\right]$$

Example 4.6 Express the inputs of the sub-channels $W_{N/2}$, i.e., $s_i, i = 1, \ldots, N$, in terms of the inputs $u_i, i = 1, \ldots, N$ of the combined channel W_N in Fig. 4.7.

Solution 4.6 From Fig. 4.7, we can write the $s_1^N = [s_1 s_2 \ldots s_N]$ vector as

$$s_1^N = \left[u_1 \oplus u_2\ u_2\ u_3 \oplus u_4\ u_4 \ldots u_{N-1} \oplus u_N\ u_N\right]$$

The inputs of the upper $W_{N/2}$ channel are

$$\left[s_1\ s_3\ s_5 \ldots s_{N-1}\right]$$

which can be explicitly written as

$$\left[u_1 \oplus u_2\ u_3 \oplus u_4 \ldots u_{N-1} \oplus u_N\right]$$

which can be expressed as

$$u_{1,o}^N \oplus u_{1,e}^N.$$

The inputs of the lower $W_{N/2}$ channel are

$$[s_2\ s_4\ s_6 \ldots s_N]$$

which can be explicitly written as

$$[u_2\ u_4 \ldots u_N]$$

which can be expressed as

$$u_{1,e}^N.$$

Thus, the inputs of the upper half channel $W_{N/2}$ are

$$u_{1,o}^N \oplus u_{1,e}^N$$

and the inputs of the lower half channel $W_{N/2}$ are

$$u_{1,e}^N.$$

Note the similarity between the code bits $u_1 \oplus u_2$, u_2 obtained for $N = 2$ and $u_{1,o}^N \oplus u_{1,e}^N, u_{1,e}^N$.

Example 4.7 Express $W_N(y_1^N|u_1^N) = Prob(y_1^N|u_1^N)$ is terms of $W_{N/2}(\cdot)$ using the general structure of Fig. 4.7.

Solution 4.7 From Fig. 4.7, it is obvious that the conditional probability expression

$$W_N(y_1^N|u_1^N) = Prob(y_1^N|u_1^N)$$

equals to

$$Prob(y_1^N|u_1^N) = Prob(y_1^N|s_1^N)$$

which can be written as

$$Prob(y_1^N|u_1^N) = Prob(y_1^{N/2}, y_{N/2+1}^N|s_{1,o}^N, s_{1,e}^N)$$

where employing the independence of $s_{1,o}^N$ from $s_{1,e}^N$, we get

$$Prob(y_1^{N/2}|s_{1,o}^N)Prob(y_{N/2+1}^N|s_{1,e}^N)$$

that is

$$Prob\left(y_1^N|u_1^N\right) = Prob\left(y_1^{N/2}|s_{1,o}^N\right)Prob(y_{N/2+1}^N|s_{1,e}^N)$$

which can also be written as

$$W_N(y_1^N|u_1^N) = W_{N/2}(y_1^{N/2}|s_{1,o}^N)W_{N/2}(y_{N/2+1}^N|s_{1,e}^N) \tag{4.15}$$

in which substituting

$$s_{1,o}^N = u_{1,o}^N \oplus u_{1,e}^N \quad s_{1,e}^N = u_{1,e}^N \tag{4.16}$$

we get

$$W_N(y_1^N|u_1^N) = W_{N/2}(y_1^{N/2}|u_{1,o}^N \oplus u_{1,e}^N)W_{N/2}(y_{N/2+1}^N|u_{1,e}^N). \tag{4.17}$$

As an example, for $N = 2$, (4.17) takes the form of

$$W_2(y_1, y_2|u_1, u_2) = W_1(y_1|u_1 \oplus u_2)W_1(y_2|u_2).$$

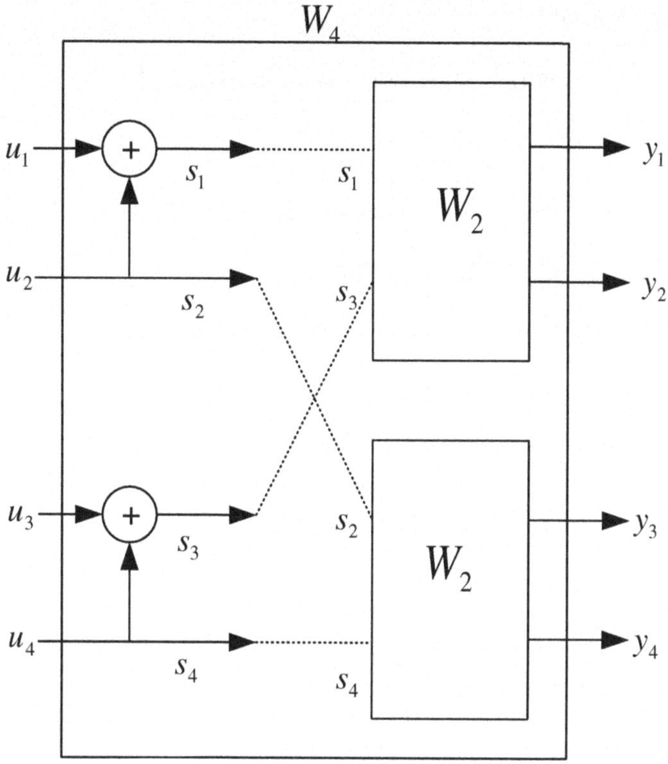

Fig. 4.8 W_4 channel

Example 4.8 For the communication channel shown in Fig. 4.8, show that

$$Prob(u_1, u_2, u_3, u_4, y_1, y_2, y_3, y_4) = Prob(s_1, s_3, y_1, y_2)Prob(s_2, s_4, y_3, y_4) \tag{4.18}$$

which can also be written as

$$W_4(u_1^4, y_1^4) = W_2(s_1, s_3, y_1^2)W_2(s_2, s_4, y_3^4). \tag{4.19}$$

Solution 4.8 Considering Fig. 4.8, we can write

$$Prob(u_1, u_2, u_3, u_4, y_1, y_2, y_3, y_4) = Prob(s_1, s_2, s_3, s_4, y_1, y_2, y_3, y_4)$$

where the right hand side can be written as in

$$Prob(s_1, s_2, s_3, s_4, y_1, y_2, y_3, y_4) = Prob(s_1, s_3, y_1, y_2)Prob(s_2, s_4, y_3, y_4).$$

Thus, we obtained

$$W_4(u_1^4, y_1^4) = W_2(s_1, s_3, y_1^2)W_2(s_2, s_4, y_3^4) \tag{4.20}$$

leading to

$$W_4(y_1^4|u_1^4) = W_2(y_1^2|u_1 \oplus u_2, u_3 \oplus u_4)W_2(y_3^4|u_2, u_4). \tag{4.21}$$

Example 4.9 Write $W_8(\cdot)$ in terms of $W_2(\cdot)$.

Solution 4.9 Considering the general recursive expression

$$W_N(y_1^N|u_1^N) = W_{N/2}(y_1^{N/2}|s_{1,o}^N)W_{N/2}(y_{N/2+1}^N|s_{1,e}^N) \tag{4.22}$$

where

$$s_{1,o}^N = u_{1,o}^N \oplus u_{1,e}^N \oplus s_{1,e}^N = u_{1,e}^N \tag{4.23}$$

For $N = 8$, we have

$$\begin{aligned} s_{1,o}^8 &= [s_1\ s_3\ s_5\ s_7] \rightarrow s_{1,o}^8 = [u_1 \oplus u_2\ u_3 \oplus u_4\ u_5 \oplus u_6\ u_7 \oplus u_8] \\ s_{1,e}^8 &= [s_2\ s_4\ s_6\ s_8] \rightarrow s_{1,e}^8 = [u_2\ u_4\ u_6\ u_8] \end{aligned} \tag{4.24}$$

and $W_8(u_1^8, y_1^8)$ can be written, employing the recursion $W_8(\cdot) = W_4(\cdot)W_4(\cdot)$, as

$$W_8(u_1^8, y_1^8) = W_4(s_1, s_3, s_5, s_7, y_1^4)W_4(s_2, s_4, s_6, s_8, y_5^8)$$

which is equal to

$$W_8(u_1^8, y_1^8) = W_4(u_1 \oplus u_2, u_3 \oplus u_4, u_5 \oplus u_6, u_7 \oplus u_8, y_1^4)W_4(u_2, u_4, u_6, u_8, y_5^8) \tag{4.25}$$

which can be written in short as

$$W_8(u_1^8, y_1^8) = W_4(u_{1,o}^8 \oplus u_{1,e}^8, y_1^4)W_4(u_{1,e}^8, y_5^8) \tag{4.26}$$

from which, we can easily derive

$$W_8(y_1^8|u_1^8) = W_4(y_1^4|u_{1,o}^N \oplus u_{1,e}^N)W_4(y_5^8|u_{1,e}^N). \tag{4.27}$$

Employing the recursion $W_4(\cdot) = W_2(\cdot)W_2(\cdot)$ for (4.26) as

$$W_4(t_1^4, r_1^4) = W_2(t_{1,o}^4 \oplus t_{1,e}^4, r_1^2)W_2(t_{1,e}^4, r_3^4)$$

where

$$t_1^4 = [u_1 \oplus u_2, u_3 \oplus u_4, u_5 \oplus u_6, u_7 \oplus u_8] \quad r_1^4 = y_1^4$$

for the first half, i.e., for $W_4(u_{1,o}^8 \oplus u_{1,e}^8, y_1^4)$, and they are

$$t_1^4 = [u_2, u_4, u_6, u_8] \quad r_1^4 = y_5^8$$

for the second half, i.e., for $W_4(u_{1,e}^8, y_5^8)$, we get

$$W_4(u_{1,o}^8 \oplus u_{1,e}^8, y_1^4) = W_2(u_1 \oplus u_2 \oplus u_5 \oplus u_6, u_3 \oplus u_4 \oplus u_7 \oplus u_8, y_1^2)W_2(u_3 \oplus u_4, u_7 \oplus u_8, y_3^4)$$
$$W_4(u_{1,e}^8, y_5^8) = W_2(u_2 \oplus u_6, u_4 \oplus u_8, y_5^6)W_2(u_4, u_6, y_7^8) \tag{4.28}$$

where we can employ the property

$$W_2(y_1^2|x_1^2) = W(y_1|u_1 \oplus u_2)W(y_2|u_2) \tag{4.29}$$

or

$$W_2(y_1^2, u_1 \oplus u_2, u_2) = W(y_1, u_1 \oplus u_2)W(y_2, u_2) \tag{4.30}$$

to further simplify (4.28).

Exercise *For $N = 16$, write the joint probability mass function between channel inputs and channel outputs of the combined channel W_{16} in terms of the input-output joint probability mass functions of the sub-combined channels W_8.*

4.2 Channel Splitting and Decoding of Polar Codes

Now let's consider the decoding operation for the combined channel of Fig. 4.4. Using the received symbols $y_1^N = \left[y_1\ y_2 \ldots y_{N-1}\ y_N\right]$, we can decode the bit u_1. For this purpose, let's define the channel W_N^1 as

$$W_N^1 : u_1 \rightarrow y_1^N. \tag{4.31}$$

After decoding u_1, we can consider the decoding of u_2, and for the decoding of u_2, we can use the previously decoded bit value of u_1, and the received symbols y_1^N. Let's define the channel W_N^2 as

$$W_N^2 : u_2 \rightarrow y_1^N, \hat{u}_1. \tag{4.32}$$

After deciding on the value of u_2, we can consider the decoding of u_3, and for this purpose we can use the received signal y_1^N and previously decided bits u_1, u_2. Let's define the channel W_N^3 as

$$W_N^3 : u_3 \rightarrow y_1^N, \hat{u}_1, \hat{u}_2. \tag{4.33}$$

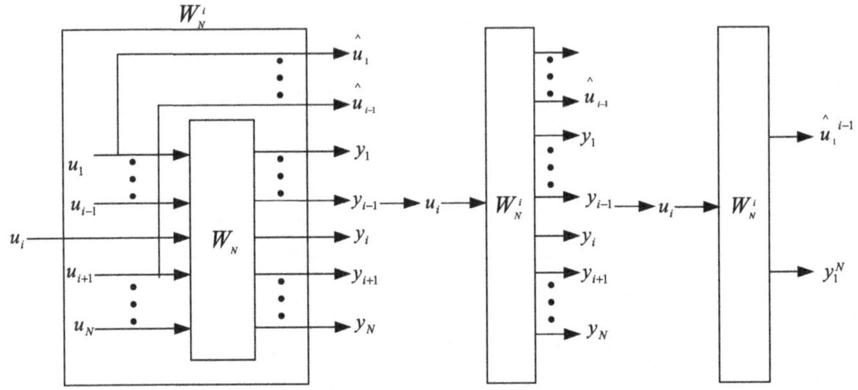

Fig. 4.9 Channel splitting and the split channel W_N^i

We can generalize this situation for the decoding of bit u_i, and define the channel W_N^i as

$$W_N^i : u_i \rightarrow y_1^N, \hat{u}_1, \hat{u}_2, \ldots, \hat{u}_{i-1} \tag{4.34}$$

which can also be written as

$$W_N^i : u_i \rightarrow y_1^N, \hat{u}_1^{i-1}. \tag{4.35}$$

The channel W_N^i can be graphically illustrated as in Fig. 4.9.

Example 4.10 Explain the meaning of

$$W_8^5 : u_5 \rightarrow y_1, y_2, y_3, y_4, y_5, y_6, y_7, y_8, \hat{u}_1, \hat{u}_2, \hat{u}_3, \hat{u}_4 \tag{4.36}$$

which can be written in a more compact manner as

$$W_8^5 : u_5 \rightarrow y_1^8, \hat{u}_1^4. \tag{4.37}$$

Solution 4.10 The expression

$$W_8^5 : u_5 \rightarrow y_1^8, \hat{u}_1^4$$

implies that, a data vector is encoded using a polar encoder and a code-word containing 8-bits is generated and transmitted through a discrete channel, and at the receiver side the signal sequence y_1^8 is available. We started to decode the data bits at the receiver side in a sequential manner, and decoded the first 4 data bits, i.e., the values of u_1, u_2, u_3 and u_4 are decided. We want to decode the data bit u_5, and we want to use all the available information for the decoding operation, for this reason, we

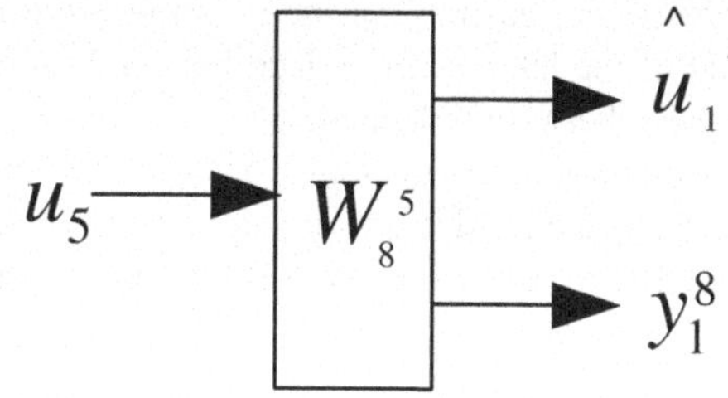

Fig. 4.10 The split channel W_8^5

consider the channel with input u_5 and outputs $y_1^8, \hat{u}_1^4$. Since we have the channel output information, using this information we can decode the data bit u_5. This is graphically depicted in Fig. 4.10.

Exercise *Explain the meaning of* W_{16}^7.

Example 4.11 Assume that the code-word length for a polar code is $N = 8$. Write all the split channels to be considered for decoding operation at the receiver side.

Solution 4.11 For $N = 8$, we can write the split channels to be used for the decoding operation of the bit at the channel input explicitly as

$$W_8^1 : u_1 \rightarrow y_1, y_2, y_3, y_4, y_5, y_6, y_7, y_8$$
$$W_8^2 : u_2 \rightarrow y_1, y_2, y_3, y_4, y_5, y_6, y_7, y_8, \hat{u}_1$$
$$W_8^3 : u_3 \rightarrow y_1, y_2, y_3, y_4, y_5, y_6, y_7, y_8, \hat{u}_1, \hat{u}_2$$
$$W_8^4 : u_4 \rightarrow y_1, y_2, y_3, y_4, y_5, y_6, y_7, y_8, \hat{u}_1, \hat{u}_2, \hat{u}_3$$
$$W_8^5 : u_5 \rightarrow y_1, y_2, y_3, y_4, y_5, y_6, y_7, y_8, \hat{u}_1, \hat{u}_2, \hat{u}_3, \hat{u}_4$$
$$W_8^6 : u_6 \rightarrow y_1, y_2, y_3, y_4, y_5, y_6, y_7, y_8, \hat{u}_1, \hat{u}_2, \hat{u}_3, \hat{u}_4, \hat{u}_5$$
$$W_8^7 : u_7 \rightarrow y_1, y_2, y_3, y_4, y_5, y_6, y_7, y_8, \hat{u}_1, \hat{u}_2, \hat{u}_3, \hat{u}_4, \hat{u}_5, \hat{u}_6$$
$$W_8^8 : u_8 \rightarrow y_1, y_2, y_3, y_4, y_5, y_6, y_7, y_8, \hat{u}_1, \hat{u}_2, \hat{u}_3, \hat{u}_4, \hat{u}_5, \hat{u}_6, \hat{u}_7$$

Example 4.12 Assume that the code-word length for a polar code is $N = 1024$. Sort the capacities of the following split channels

$$W_{1024}^{64} \quad W_{1024}^{128} \quad W_{1024}^{256} \quad W_{1024}^{512}$$

in ascending order.

Solution 4.12 We can sort the capacities of the given split channels as

$$C\left(W_{1024}^{64}\right) < C\left(W_{1024}^{128}\right) < C\left(W_{1024}^{256}\right) < C\left(W_{1024}^{512}\right).$$

Note: For easiness of the notation, from this point on, we will not use ^ on the head of the letters of the decoded symbols, i.e., we will use u_i for the place of $\hat{u}_i$ unless otherwise indicated.

Up to now, have seen the mathematical expressions W_N which implies the combined channel, and W_N^i which implies the ith split channel. In the subsequent sections we will use a new expression W^N which implies N parallel W channels whose inputs are $x_1, x_2, \ldots, x_N$.

4.3 Mathematical Description of Successive Cancellation Algorithm

Consider the split channel

$$W_N^i : u_i \rightarrow y_1^N, u_1^{i-1} \tag{4.38}$$

where u_i is the channel input, y_1^N, u_1^{i-1} are the channel outputs, and using the channel outputs, we want to determine the channel input value, i.e., we want to find whether $u_i = 0$ or $u_i = 1$. We can achieve it considering the following proposition.

Proposition P4.1 *For the split channel W_N^i, the conditional input probability can be calculated using*

$$W_N^i : Prob(y_1^N, u_1^{i-1}|u_i) = \frac{1}{2^{N-1}} \sum_{u_{i+1}^N} Prob(y_1^N|u_1^N) \tag{4.39}$$

which is equal to

$$W_N^i : Prob(y_1^N, u_1^{i-1}|u_i) = \frac{1}{2^{N-1}} \sum_{u_{i+1}^N} Prob(y_1^N|x_1^N) \tag{4.40}$$

The Eqs. (4.39) and (4.40) can also be written as

$$W_N^i(y_1^N, u_1^{i-1}|u_i) = \frac{1}{2^{N-1}} \sum_{u_{i+1}^N} W_N(y_1^N|u_1^N) \tag{4.41}$$

which is equals to

$$W_N^i(y_1^N, u_1^{i-1}|u_i) = \frac{1}{2^{N-1}} \sum_{u_{i+1}^N} W^N(y_1^N|x_1^N) \tag{4.42}$$

Note also that $W^N\left(y_1^N|x_1^N\right) = W(y_1|x_1)W(y_2|x_2)\ldots W(y_N|x_N)$.

Proof P4.1 To determine whether $u_i = 0$ or $u_i = 1$ is transmitted, we calculate the probabilities

$$Prob(u_i = 0, y_1^N, u_1^{i-1}) \quad Prob(u_i = 1, y_1^N, u_1^{i-1})$$

and compare them to each other to determine the transmitted bit as

$$Prob(u_i = 0, y_1^N, u_1^{i-1}) \overbrace{\underbrace{\substack{\geq \\ <}}_{u_i=1}}^{u_i=0} Prob(u_i = 1, y_1^N, u_1^{i-1}). \tag{4.43}$$

The inequality (4.43) can be written in a more compact manner as

$$\frac{Prob(u_i = 0, y_1^N, u_1^{i-1})}{Prob(u_i = 1, y_1^N, u_1^{i-1})} \overbrace{\underbrace{\substack{\geq \\ <}}_{u_i=1}}^{u_i=0} 1$$

where using the probability property $Prob(a, b) = Prob(b|a)Prob(a)$, we get

$$\frac{Prob\big(y_1^N, u_1^{i-1}|u_i = 0\big)Prob(u_i = 0)}{Prob\big(y_1^N, u_1^{i-1}|u_i = 1\big)Prob(u_i = 1)} \overbrace{\underbrace{\substack{\geq \\ <}}_{u_i=1}}^{u_i=0} 1$$

in which employing

$$Prob(u_i = 0) = Prob(u_i = 1) = \frac{1}{2}$$

we get

$$\frac{Prob(y_1^N, u_1^{i-1}|u_i = 0)}{Prob(y_1^N, u_1^{i-1}|u_i = 1)} \overbrace{\underbrace{\substack{\geq \\ <}}_{u_i=1}}^{u_i=0} 1$$

The decoding logic can be summarized as

$$u_i = \begin{cases} 0 \text{ if } \frac{Prob(y_1^N, u_1^{i-1}|u_i = 0)}{Prob(y_1^N, u_1^{i-1}|u_i = 1)} \geq 1 \text{or } u_i \text{ is a frozen bit} \\ 1 \text{ otherwise.} \end{cases} \tag{4.44}$$

Now, let's consider how to evaluate the conditional probabilities in (4.44) Since the formulas we generate are similar for $u_i = 0$ and $u_i = 1$, for the moment, let's use only u_i without considering its value. We can write the probability

$$Prob(y_1^N, u_1^{i-1}|u_i)$$

as

$$Prob(y_1^N, u_1^{i-1}|u_i) = \frac{Prob(u_i, y_1^N, u_1^{i-1})}{Prob(u_i)} = \frac{Prob(u_1^i, y_1^N)}{Prob(u_i)}$$

where for the numerator term employing the probability property

$$p(w, x) = \sum_{y,z} p(w, x, y, z)$$

we obtain

$$Prob(y_1^N, u_1^{i-1}|u_i) = \frac{\sum_{u_{i+1}^N} Prob(u_1^N, y_1^N)}{Prob(u_i)}$$

where employing

$$Prob(u_1^N, y_1^N) = Prob(y_1^N|u_1^N)Prob(u_1^N)$$

we get

$$Prob(y_1^N, u_1^{i-1}|u_i) = \frac{\sum_{u_{i+1}^N} Prob(y_1^N|u_1^N)Prob(u_1^N)}{Prob(u_i)}. \tag{4.45}$$

Since

$$Prob(u_i) = \frac{1}{2} \tag{4.46}$$

then $Prob(u_1^N)$ can be calculated as

$$\begin{aligned}
&Prob(u_1^N) = Prob(u_1, u_2, \ldots, u_{N-1}, u_N) \rightarrow \\
&Prob(u_1^N) = Prob(u_1)Prob(u_2)\ldots Prob(u_N) \rightarrow \\
&Prob(u_1^N) = \frac{1}{2^N}.
\end{aligned} \tag{4.47}$$

Substituting (4.47) and (4.46) into (4.45), we get

$$Prob(y_1^N, u_1^{i-1}|u_i) = \frac{\sum_{u_{i+1}^N} Prob(y_1^N|u_1^N)\frac{1}{2^N}}{\frac{1}{2}}. \quad (4.48)$$

leading to

$$W_N^i: \quad Prob(y_1^N, u_1^{i-1}|u_i) = \frac{1}{2^{N-1}} \sum_{u_{i+1}^N} Prob(y_1^N|u_1^N). \quad (4.49)$$

The probability expression $Prob(y_1^N|u_1^N)$ in (4.49) can be calculated as

$$Prob(y_1^N|u_1^N) = Prob(y_1^N|x_1^N) \quad (4.50)$$

where

$$x_1^N = u_1^N G. \quad (4.51)$$

For the probability expression

$$Prob(y_1^N|x_1^N)$$

we have

$$Prob(y_1^N|x_1^N) = \prod_{i=1}^{N} Prob(y_i|x_i) \quad (4.52)$$

Example 4.13 For $N = 2$, indicate the combined and split channels and, write the conditional probability expressions for the inputs of the split channels.

Solution 4.13 The combined channel W_2 has two inputs u_1, u_2 and two outputs y_1, y_2 and it is shown in Fig. 4.11.

The split channels are written as

$$\begin{aligned} W_2^1 &: u_1 \rightarrow y_1, y_2 \\ W_2^2 &: u_2 \rightarrow y_1, y_2, u_1 \end{aligned} \quad (4.53)$$

and they are graphically depicted as in Fig. 4.12.

When conditional probability expression

$$W_N^i: \quad Prob(y_1^N, u_1^{i-1}|u_i) = \frac{1}{2^{N-1}} \sum_{u_{i+1}^N} Prob(y_1^N|u_1^N) \quad (4.54)$$

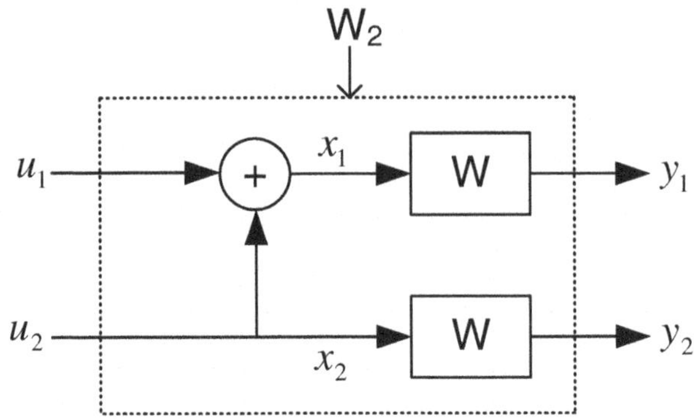

Fig. 4.11 W_2 channel for Example 4.13

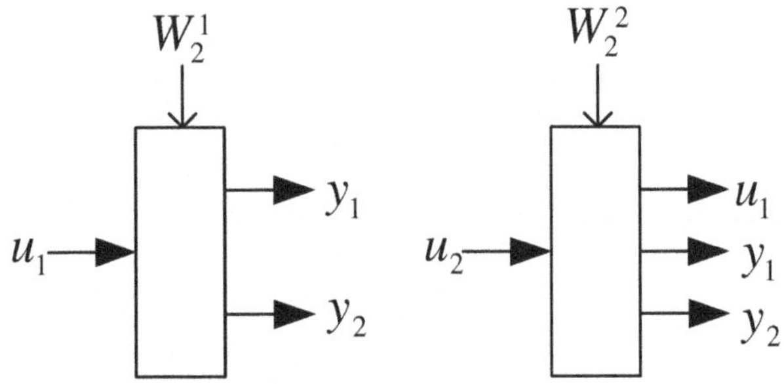

Fig. 4.12 The split channels W_2^1 and W_2^2

is expanded for $N = 2$ and $i = 1$, i.e., for the split channel W_2^1, we get

$$Prob(y_1^2, u_1^0|u_1) = \frac{1}{2^{2-1}} \sum_{u_2^2} Prob(y_1^2|u_1^2)$$

where using $\hat{u}_1^0 = \{\}$ and $u_2^2 = u_2$, we get

$$Prob(y_1^2|u_1) = \frac{1}{2} \sum_{u_2} Prob(y_1^2|u_1^2)$$

which can be written as

$$Prob(y_1^2|u_1) = \frac{1}{2} \sum_{u_2} Prob(y_1|u_1)Prob(y_2|u_1)$$

in which employing

$$Prob(y_i|u_i) = Prob(y_i|x_i)$$

we obtain

$$Prob(y_1^2|u_1) = \frac{1}{2}\sum_{u_2} Prob(y_1|x_1)Prob(y_2|x_2)$$

where substituting

$$x_1 = u_1 \oplus u_2 \quad x_2 = u_2$$

we obtain

$$Prob(y_1^2|u_1) = \frac{1}{2}\sum_{u_2} Prob(y_1|u_1 \oplus u_2)Prob(y_2|u_2) \tag{4.55}$$

If (4.55) is evaluated for $u_1 = 0$, we get

$$Prob(y_1^2|u_1 = 0) = \frac{1}{2}\sum_{u_2} Prob(y_1|0 \oplus u_2)Prob(y_2|u_2). \tag{4.56}$$

Expanding the summation in (4.56) for $u_2 = 0$ and $u_2 = 1$, we get

$$Prob(y_1^2|u_1 = 0) = \frac{1}{2}(Prob(y_1|0 \oplus 0)Prob(y_2|0) + Prob(y_1|0 \oplus 1)Prob(y_2|1))$$

which can be simplified as

$$Prob(y_1^2|u_1 = 0) = \frac{1}{2}(Prob(y_1|0)Prob(y_2|0) + Prob(y_1|1)Prob(y_2|1))$$

which can also be written as

$$W_2^1(y_1^2|u_1 = 0) = \frac{1}{2}(W(y_1|0)W(y_2|0) + W(y_1|1)W(y_2|1)).$$

In a similar manner, if (4.55) is evaluated for $u_1 = 1$, we get

$$Prob(y_1^2|u_1 = 1) = \frac{1}{2}\sum_{u_2} Prob(y_1|1 \oplus u_2)Prob(y_2|u_2). \tag{4.57}$$

Expanding the summation in (4.57) for $u_2 = 0$ and $u_2 = 1$, we get

$$Prob(y_1^2|u_1 = 1) = \frac{1}{2}(Prob(y_1|1 \oplus 0)Prob(y_2|0) + Prob(y_1|1 \oplus 1)Prob(y_2|1))$$

which can be simplified as

$$Prob(y_1^2|u_1 = 1) = \frac{1}{2}(Prob(y_1|1)Prob(y_2|0) + Prob(y_1|0)Prob(y_2|1))$$

which can also be written as

$$W_2^1(y_1^2|u_1 = 1) = \frac{1}{2}(W(y_1|1)W(y_2|0) + W(y_1|0)W(y_2|1)).$$

Hence, for the channel W_2^1, we have the conditional probabilities

$$W_2^1: \begin{array}{l} Prob(y_1^2|u_1 = 0) = \frac{1}{2}(Prob(y_1|0)Prob(y_2|0) + Prob(y_1|1)Prob(y_2|1)) \\ Prob(y_1^2|u_1 = 1) = \frac{1}{2}(Prob(y_1|1)Prob(y_2|0) + Prob(y_1|0)Prob(y_2|1)). \end{array} \tag{4.58}$$

To determine the value of the channel input u_1 we compare the probabilities $Prob(y_1^2|u_1 = 0)$ and $Prob(y_1^2|u_1 = 1)$ to each other, or compare their ratio to 1as in

$$\frac{Prob(y_1^2|u_1 = 0)}{Prob(y_1^2|u_1 = 1)} \overbrace{\underbrace{\gtrless}_{u_i=1}}^{u_i=0} 1.$$

When conditional probability expression

$$W_N^i: \quad Prob(y_1^N, u_1^{i-1}|u_i) = \frac{1}{2^{N-1}} \sum_{u_{i+1}^N} Prob(y_1^N|u_1^N) \tag{4.59}$$

is evaluated for $N = 2$ and $i = 2$, i.e., evaluated for the split channel W_2^2, we get

$$Prob(y_1^2, u_1^1|u_2) = \frac{1}{2^{2-1}} \sum_{u_3^2} Prob(y_1^2|u_1^2) \tag{4.60}$$

where using $u_3^2 = \{\}$, we get

$$Prob(y_1^2, u_1|u_2) = \frac{1}{2} Prob(y_1^2|u_1^2)$$

which can be written as

$$Prob(y_1^2, u_1|u_2) = \frac{1}{2} Prob(y_1|u_1)Prob(y_2|u_1)$$

in which employing

$$Prob(y_i|u_i) = Prob(y_i|x_i)$$

we obtain

$$Prob(y_1^2, u_1|u_2) = \frac{1}{2}Prob(y_1|x_1)Prob(y_2|x_2)$$

where substituting

$$x_1 = u_1 \oplus u_2 \quad x_2 = u_2$$

we obtain

$$Prob(y_1^2, u_1|u_2) = \frac{1}{2}Prob(y_1|u_1 \oplus u_2)Prob(y_2|u_2) \tag{4.61}$$

where u_1 is the value of input bit of the split channel W_2^1 evaluated in the previous stage.

When (4.61) is evaluated for $u_2 = 0$, we get

$$Prob(y_1^2, u_1|u_2 = 0) = \frac{1}{2}Prob(y_1|u_1 \oplus 0)Prob(y_2|0)$$

leading to

$$Prob(y_1^2, u_1|u_2 = 0) = \frac{1}{2}Prob(y_1|u_1)Prob(y_2|0)$$

which can also be written as

$$W_2^2(y_1^2, u_1|u_2 = 0) = \frac{1}{2}W(y_1|u_1)W(y_2|0).$$

In a similar manner, when (4.61) is evaluated for $u_2 = 1$, we get

$$Prob(y_1^2, u_1|u_2 = 1) = \frac{1}{2}Prob(y_1|u_1 \oplus 1)Prob(y_2|1)$$

leading to

$$Prob(y_1^2, u_1|u_2 = 1) = \frac{1}{2}Prob(y_1|u_1')Prob(y_2|1)$$

which can also be written as

$$W_2^2(y_1^2, u_1|u_2 = 1) = \frac{1}{2}W(y_1|u_1')W(y_2|1)$$

Hence, for the channel W_2^2, we have the conditional probabilities

$$W_2^1: \begin{matrix} Prob(y_1^2, u_1|u_2 = 0) = \frac{1}{2}Prob(y_1|u_1)Prob(y_2|0) \\ Prob(y_1^2, u_1|u_2 = 1) = \frac{1}{2}Prob(y_1|u_1')Prob(y_2|1). \end{matrix}$$

To determine the value of the channel input u_2 we compare the probabilities $Prob(y_1^2, u_1|u_2 = 0)$ and $Prob(y_1^2, u_1|u_2 = 1)$ to each other, or compare their ration to 1 as in

$$\frac{Prob(y_1^2, u_1|u_2 = 0)}{Prob(y_1^2, u_1|u_2 = 1)} \gtrless \overbrace{\underbrace{\geq <}_{u_2=1}}^{u_2=0} 1 \tag{4.62}$$

which can also be written as

$$\frac{W_2^2(y_1^2, u_1|u_2 = 0)}{W_2^2(y_1^2, u_1|u_2 = 1)} \gtrless \overbrace{\underbrace{\geq <}_{u_2=1}}^{u_2=0} 1. \tag{4.63}$$

Example 4.14 For $N = 4$, indicate the combined and split channels and, write the conditional probability expressions for the inputs of the split channels.

The conditional probability expression

$$W_N^i(y_1^N, u_1^{i-1}|u_i) = \frac{1}{2^{N-1}} \sum_{u_{i+1}^N} W_N(y_1^N|u_1^N) \tag{4.64}$$

for the split channel inputs is not feasible even for small or moderate values of N. For instance, for $N = 20$, the calculation of

$$\frac{1}{2^{N-1}} \sum_{u_6^{20}} W_{20}(y_1^{20}|u_1^{20}) \tag{4.65}$$

involves 2^{14} terms to be computed. For this reason,

$$W_N^i(y_1^N, u_1^{i-1}|u_i) = \frac{1}{2^{N-1}} \sum_{u_{i+1}^N} W_N(y_1^N|u_1^N) \tag{4.66}$$

is not used for decoding purposes. A fast and recursive approach is presented in [1] for the recursive successive cancelation decoding of polar codes. In the next section, we will explain the recursive decoding approach which is a feasible and practical approach for the decoding of polar codes.

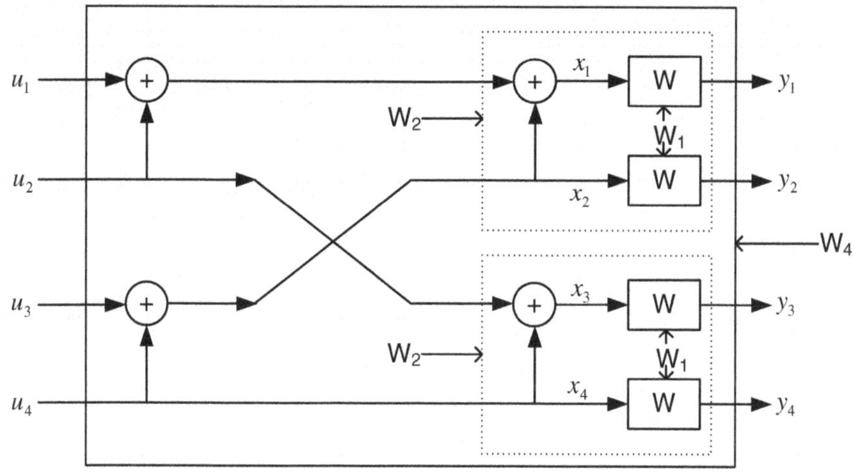

Fig. 4.13 W_4 and W_2 channels

4.4 Recursive Computation of the Successive Cancellation Algorithm

The successive decoding method presented in the previous section is not feasible even for moderate length code-words. For this reason, Arıkan in its original paper [1] proposed the recursive computation of the successive cancellation decoding of polar codes. Before deriving the formulas necessary for the recursive decoding operation, let's provide some information about recursive channel transformation.

4.4.1 Recursive Channel Transformation

Let's consider the polar encoder structure, constructed for $N = 4$, shown in Fig. 4.13. If the encoding operation of the polar codes is followed from right to left, we see that the channels are combined in a similar manner till the beginning of encoding operation.

Now let's consider the split channels. For this reason, for easiness of the illustration let's consider the split channels W_2^1, W_4^1, and W_4^2. The split channels W_2^1 are shown in Fig. 4.14.

The split channel W_4^1 is shown in Fig. 4.15 where it is seen that W_4^1 is formed combining two split W_2^1 channels, i.e., we can write that

$$(W_2^1, W_2^1) \rightarrow W_4^1 \tag{4.67}$$

where we have

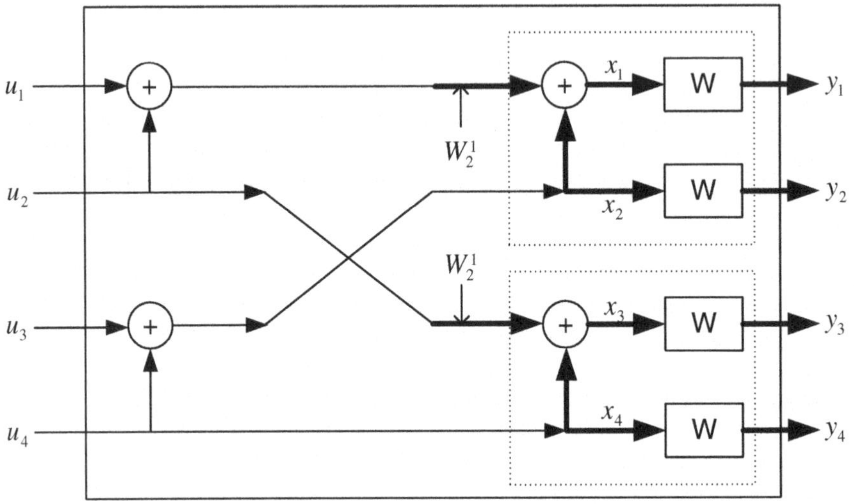

Fig. 4.14 The split channels W_2^1

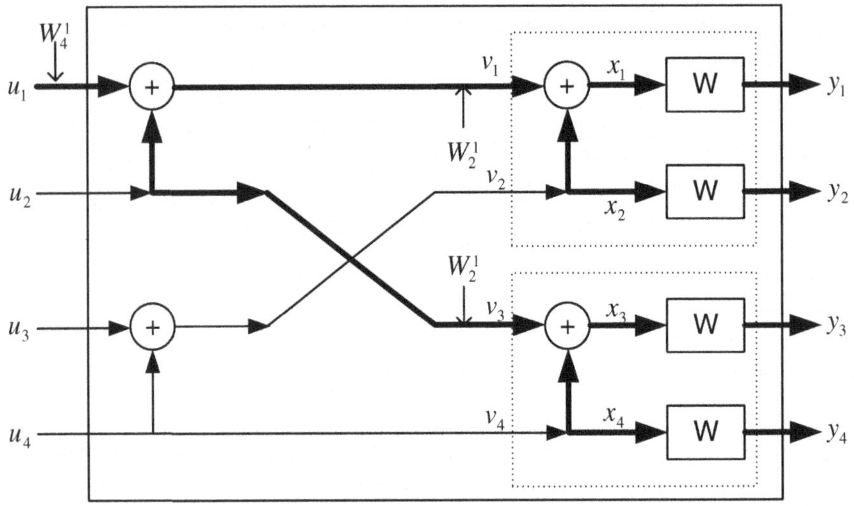

Fig. 4.15 The split channel W_4^1

$$\begin{aligned} W_2^1\text{up}: &\quad v_1 \to y_1, y_2 \\ W_2^1\text{down}: &\quad v_3 \to y_3, y_4 \\ W_4^1: &\quad u_1 \to y_1, y_2, y_3, y_4 \end{aligned} \tag{4.68}$$

The split channel W_4^1 is depicted in Fig. 4.16 alone.

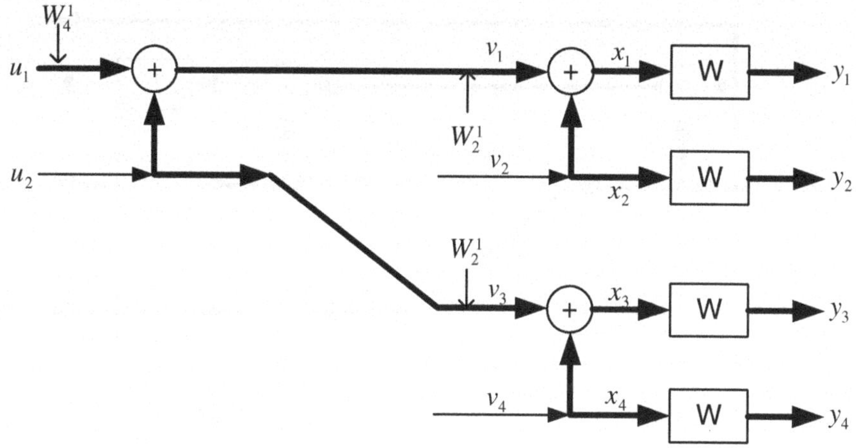

Fig. 4.16 The path of the split channel W_4^1

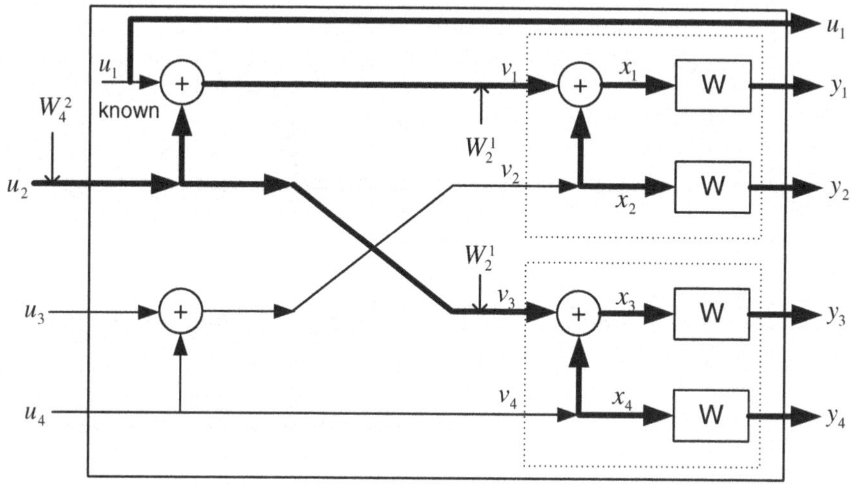

Fig. 4.17 The split channel W_4^2

The split channel W_4^2 is shown in Fig. 4.17 by bold lines, and it is seen from Fig. 4.17 that the channel W_4^2 is constructed by combining the two split channels W_2^1, i.e., we can write that

$$(W_2^1, W_2^1) \rightarrow W_4^2 \tag{4.69}$$

where we have

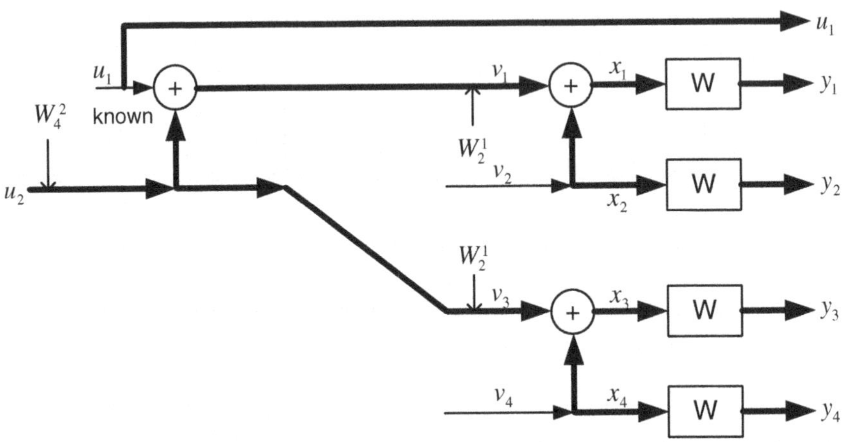

Fig. 4.18 The path of the split channel W_4^2

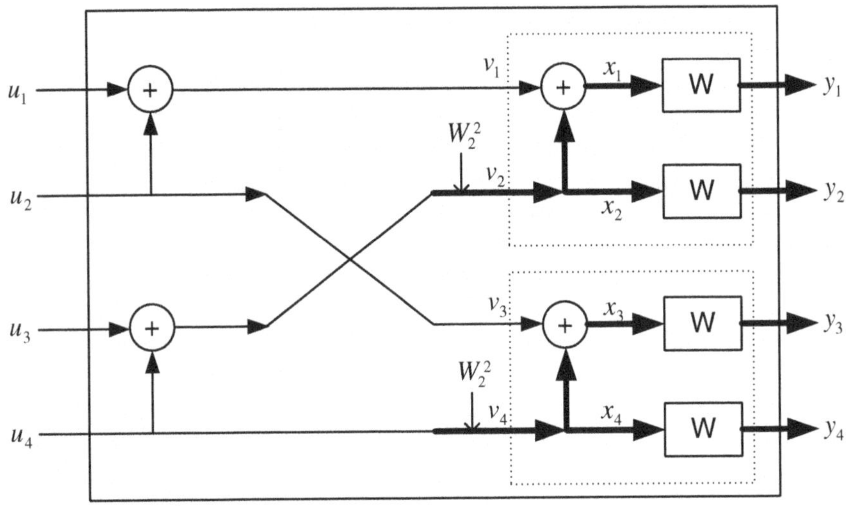

Fig. 4.19 The split channels W_2^2

$$\begin{aligned} &W_2^1\text{up}: && v_1 \to y_1, y_2 \\ &W_2^1\text{down}: && v_3 \to y_3, y_4 \\ &W_4^2: && u_1 \to y_1, y_2, y_3, y_4, u_1 \end{aligned} \tag{4.70}$$

The split channel W_4^2 is depicted in Fig. 4.18 alone.
The split channels W_2^2 are shown in Fig. 4.19.

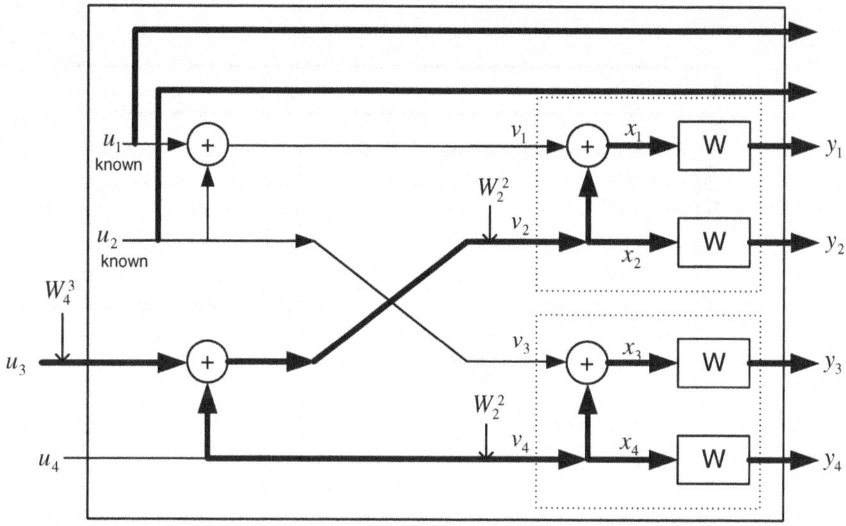

Fig. 4.20 The split channel W_4^3

The split channel W_4^3 is shown in Fig. 4.20 by bold lines, and it is seen from Fig. 4.20 that the channel W_4^3 is constructed by combining the two split channels W_2^2, i.e., we can write that

$$(W_2^2, W_2^2) \rightarrow W_4^3 \tag{4.71}$$

where we have

$$\begin{aligned} W_2^2\text{up}: \quad & v_2 \rightarrow y_1, y_2, v_1 \\ W_2^2\text{down}: \quad & v_4 \rightarrow y_3, y_4, v_3 \\ W_4^3: \quad & u_3 \rightarrow y_1, y_2, y_3, y_4, u_1, u_2 \end{aligned} \tag{4.72}$$

The split channel W_4^4 is shown in Fig. 4.21 by bold lines, and it is seen from Fig. 4.21 that the channel W_4^4 is constructed by combining the two split channels W_2^2, i.e., we can write that

$$(W_2^2, W_2^2) \rightarrow W_4^4 \tag{4.73}$$

where we have

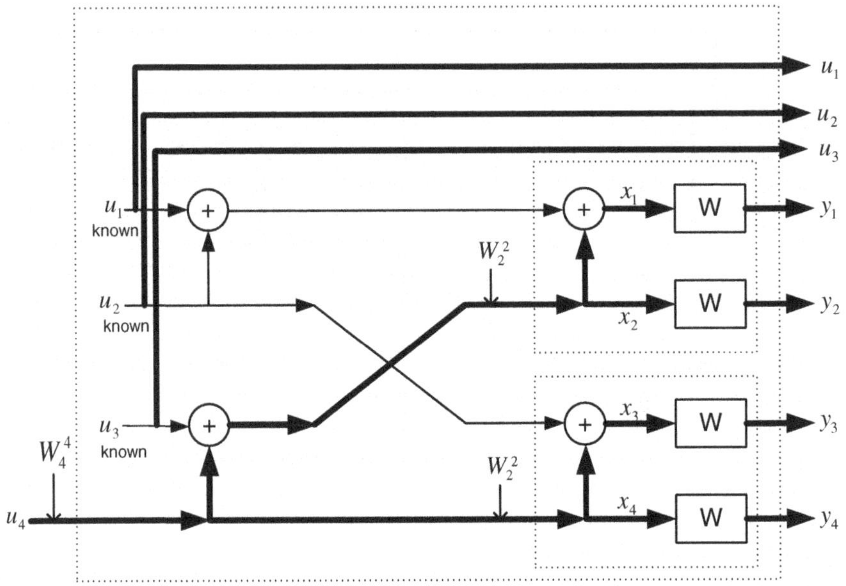

Fig. 4.21 The split channel W_4^4

$$
\begin{aligned}
&W_2^2\text{up}: && v_2 \rightarrow y_1, y_2, v_1 \\
&W_2^2\text{down}: && v_4 \rightarrow y_3, y_4, v_3 \\
&W_4^4: && u_4 \rightarrow y_1, y_2, y_3, y_4, u_1, u_2, u_3
\end{aligned}
\tag{4.74}
$$

From the previous discussion, we see that when two split channels W_N^i are combined, we get the split channels either W_{2N}^{2i-1} or W_{2N}^{2i}. This result can be expressed as

$$
(W_N^i, W_N^i) \rightarrow (W_{2N}^{2i-1}, W_{2N}^{2i}) \tag{4.75}
$$

Example 4.15 For the channel shown in Fig. 4.22, show that for the split channel W_2^1, we have

$$
\begin{aligned}
W_2^1 : Prob(y_1^2|u_1) &= \sum_{u_2} \frac{1}{2} Prob(y_1^2|u_1^2) \\
&= \sum_{u_2} \frac{1}{2} Prob(y_1|u_1 \oplus u_2) Prob(y_2|u_2)
\end{aligned}
$$

which can also be written as

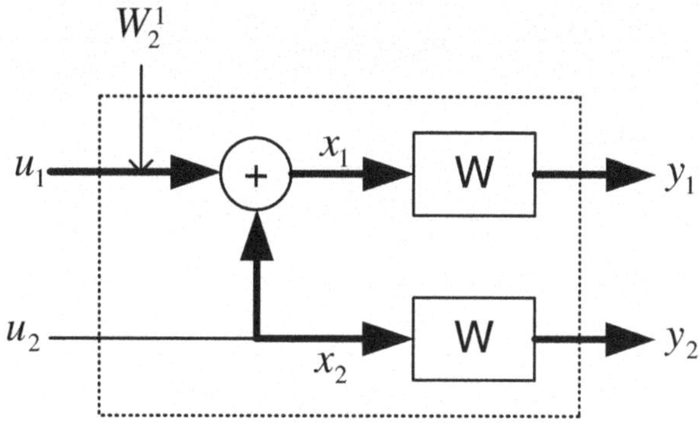

Fig. 4.22 The split channel W_2^1 for Example 4.15

$$W_2^1(y_1^2|u_1) = \sum_{u_2} \frac{1}{2} W_2(y_1^2|u_1^2)$$
$$= \sum_{u_2} \frac{1}{2} W(y_1|u_1 \oplus u_2) W(y_2|u_2) \tag{4.76}$$

and for the split channel W_2^2, we have

$$W_2^2 : Prob(y_1^2, u_1|u_2) = \frac{1}{2} Prob(y_1^2|u_1^2)$$
$$= \frac{1}{2} Prob(y_1|u_1 \oplus u_2) Prob(y_2|u_2)$$

which can also be written as

$$W_2^2(y_1^2, u_1|u_2) = \frac{1}{2} W_2(y_1^2|u_1^2)$$
$$= \frac{1}{2} W(y_1|u_1 \oplus u_2) W(y_2|u_2). \tag{4.77}$$

Solution 4.15 The conditional probability $Prob(y_1^2|u_1)$ can be written as

$$Prob(y_1^2|u_1) = \frac{Prob(y_1^2, u_1)}{Prob(u_1)}$$

in which using the property

$$p(x) = \sum_y p(x, y)$$

we get

$$Prob(y_1^2|u_1) = \frac{1}{Prob(u_1)} \sum_{u_2} Prob(y_1^2, u_1, u_2)$$

which can be written as

$$Prob(y_1^2|u_1) = \frac{1}{Prob(u_1)} \sum_{u_2} Prob(y_1^2|u_1^2)Prob(u_1)Prob(u_2)$$

where using

$$Prob(u_1) = Prob(u_2) = 1/2$$

we get

$$Prob(y_1^2|u_1) = \frac{1}{2} \sum_{u_2} Prob(y_1^2|u_1^2)$$

which can be written as

$$W_2^1(y_1^2|u_1) = \frac{1}{2} \sum_{u_2} W_2(y_1^2|u_1^2)$$

where using the property

$$W_{2N}(\cdot) = W_N(\cdot)W_N(\cdot)$$

we obtain

$$W_2^1(y_1^2|u_1) = \frac{1}{2} \sum_{u_2} W(y_1|u_1 \oplus u_2)W(y_2|u_2). \tag{4.78}$$

The probability expression

$$Prob(y_1^2, u_1|u_2)$$

can be written as

$$Prob(y_1^2, u_1|u_2) = \frac{Prob(y_1^2, u_1^2)}{Prob(u_2)} \rightarrow Prob(y_1^2, u_1|u_2) = \frac{Prob(y_1^2|u_1^2)Prob(u_1^2)}{Prob(u_2)}$$

where using

$$Prob(y_1^2|u_1^2) = Prob(y_1^2|x_1^2) \rightarrow Prob(y_1^2|u_1^2) = Prob(y_1|x_1)Prob(y_2|x_2)$$

$$Prob(u_1^2) = Prob(u_1)Prob(u_2) \text{ and } Prob(u_1) = Prob(u_2) = \frac{1}{2}$$

we obtain

$$Prob(y_1^2, u_1|u_2) = \frac{1}{2}Prob(y_1|x_1)Prob(y_2|x_2)$$

leading to

$$Prob(y_1^2, u_1|u_2) = \frac{1}{2}Prob(y_1|u_1 \oplus u_2)Prob(y_2|u_2)$$

which can be expressed also as

$$\begin{aligned} W_2^2(y_1^2, u_1|u_2) &= \frac{1}{2}W_2(y_1^2|u_1^2) \\ &= \frac{1}{2}W(y_1|u_1 \oplus u_2)W(y_2|u_2). \end{aligned} \tag{4.79}$$

Note that W_N denotes the combined channel and W_N^i indicates the split channel.

4.4.2 Butterfly Structure

The recursive channel construction logic in

$$(W_N^i, W_N^i) \rightarrow (W_{2N}^{2i-1}, W_{2N}^{2i}) \tag{4.80}$$

can be graphically illustrated using butterfly patterns. For $N = 8$, the recursive channel using butterfly patterns is illustrated in Fig. 4.23.

4.5 Recursive Calculation of Conditional Channel Input Probabilities

Let's write the open forms of the split channels W_{2N}^{2i-1} and W_{2N}^{2i} as

$$\begin{aligned} &W_{2N}^{2i-1} : u_{2i-1} \rightarrow y_1^{2N}, u_1^{2i-2} \\ &W_{2N}^{2i} : u_{2i} \rightarrow y_1^{2N}, u_1^{2i-1} \end{aligned} \tag{4.81}$$

Let's solve some examples to prepare ourselves for the derivation of the recursive expression for the calculation of conditional channel input probabilities.

$$W_N^i : Prob(y_1^N, u_1^{i-1}|u_i) = \frac{1}{2^{N-1}}\sum_{u_{i+1}^N} Prob(y_1^N|u_1^N)$$

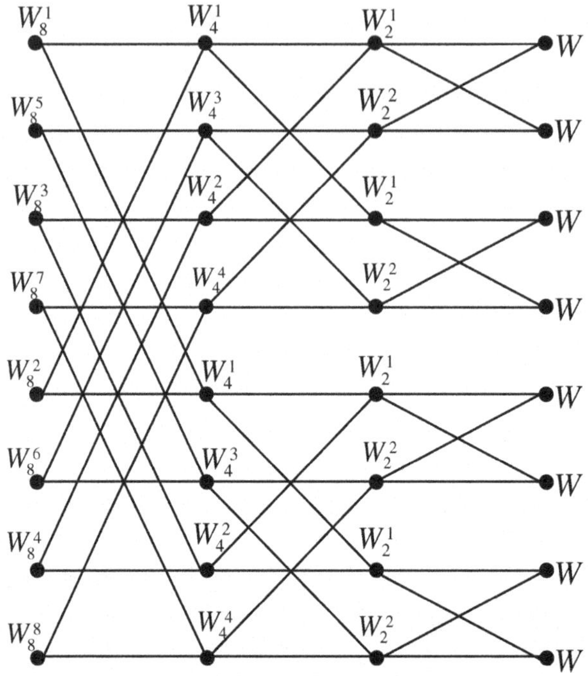

Fig. 4.23 Butterfly structure for recursive channel construction

which can also be written as

$$W_N^i(y_1^N, u_1^{i-1}|u_i) = \frac{1}{2^{N-1}} \sum_{u_{i+1}^N} W_N(y_1^N|u_1^N) \tag{4.82}$$

express

$$\sum_{u_{2i+1,e}^{2N}} \frac{1}{2^{N-1}} W_N(y_{N+1}^{2N}|u_{1,e}^{2N}) \tag{4.83}$$

by conditional input probability of a split channel.

Solution 4.16 Using the identity

$$W_N^i(y_1^N, u_1^{i-1}|u_i) = \frac{1}{2^{N-1}} \sum_{u_{i+1}^N} W_N(y_1^N|u_1^N)$$

we can express

$$\frac{1}{2^{N-1}} \sum_{u_{2i+1,e}^{2N}} W_N(y_{N+1}^{2N} | u_{1,e}^{2N})$$

as

$$W_N^i(y_{N+1}^{2N}, u_{1,e}^{2i-2} | u_{2i})$$

i.e.,

$$W_N^i\left(y_{N+1}^{2N}, u_{1,e}^{2i-2} | u_{2i}\right) = \frac{1}{2^{N-1}} \sum_{u_{2i+1,e}^{2N}} W_N(y_{N+1}^{2N} | u_{1,e}^{2N}) \tag{4.84}$$

Example 4.17 Using the identity

$$W_N^i(y_1^N, u_1^{i-1} | u_i) = \frac{1}{2^{N-1}} \sum_{u_{i+1}^N} W_N(y_1^N | u_1^N)$$

express

$$\sum_{u_{2i+1,o}^{2N}} \frac{1}{2^{N-1}} W_N(y_1^N | u_{1,o}^{2N} \oplus u_{1,e}^{2N})$$

by the conditional input probability of a split channel.

Solution 4.17 Using the identity

$$W_N^i(y_1^N, u_1^{i-1} | u_i) = \frac{1}{2^{N-1}} \sum_{u_{i+1}^N} W_N(y_1^N | u_1^N)$$

we can express

$$\sum_{u_{2i+1,o}^{2N}} \frac{1}{2^{N-1}} W_N(y_1^N | u_{1,o}^{2N} \oplus u_{1,e}^{2N})$$

as

$$W_N^i(y_1^N, u_{1,o}^{2i-2} \oplus u_{1,e}^{2i-2} | u_{2i-1} \oplus u_{2i})$$

i.e.,

$$W_N^i(y_1^N, u_{1,o}^{2i-2} \oplus u_{1,e}^{2i-2} | u_{2i-1} \oplus u_{2i}) = \sum_{u_{2i+1,o}^{2N}} \frac{1}{2^{N-1}} W_N(y_1^N | u_{1,o}^{2N} \oplus u_{1,e}^{2N}) \tag{4.85}$$

Proposition 4.1 *For the channel W_{2N}^{2i-1}, considering the channel output, the conditional probability of the channel input u_{2i-1} can be calculated recursively as*

$$Prob(y_1^{2N}, u_1^{2i-2}|u_{2i-1}) = \frac{1}{2}\sum_{u_{2i}} Prob(y_1^N, u_{1,o}^{2i-2} \oplus u_{1,e}^{2i-2}|u_{2i-1} \oplus u_{2i}) Prob(y_{N+1}^{2N}, u_{1,e}^{2i-2}|u_{2i}) \tag{4.86}$$

which can be written as

$$W_{2N}^{2i-1}(y_1^{2N}, u_1^{2i-2}|u_{2i-1}) = \frac{1}{2}\sum_{u_{2i}} W_N^i(y_1^N, u_{1,o}^{2i-2} \oplus u_{1,e}^{2i-2}|u_{2i-1} \oplus u_{2i}) W_N^i(y_{N+1}^{2N}, u_{1,e}^{2i-2}|u_{2i}) \tag{4.87}$$

and similarly for the channel W_{2N}^{2i}, considering the channel output, the conditional probability of the channel input u_{2i-1} can be calculated recursively as

$$Prob(y_1^{2N}, u_1^{2i-1}|u_{2i}) = \frac{1}{2} Prob(y_1^N, u_{1,o}^{2i-2} \oplus u_{1,e}^{2i-2}|u_{2i-1} \oplus u_{2i}) Prob(y_{N+1}^{2N}, u_{1,e}^{2i-2}|u_{2i}) \tag{4.88}$$

which can be written as

$$W_{2N}^{2i}(y_1^{2N}, u_1^{2i-1}|u_{2i}) = \frac{1}{2} W_N^i(y_1^N, u_{1,o}^{2i-2} \oplus u_{1,e}^{2i-2}|u_{2i-1} \oplus u_{2i}) W_N^i(y_{N+1}^{2N}, u_{1,e}^{2i-2}|u_{2i}). \tag{4.89}$$

Proof 4.1 In fact, in Example 4.15, we showed that

$$\begin{aligned} W_2^1(y_1^2|u_1) &= \sum_{u_2} \frac{1}{2} W_2(y_1^2|u_1^2) \\ &= \sum_{u_2} \frac{1}{2} W(y_1|u_1 \oplus u_2) W(y_2|u_2) \end{aligned}$$

and

$$\begin{aligned} W_2^2(y_1^2, u_1|u_2) &= \frac{1}{2} W_2(y_1^2|u_1^2) \\ &= \frac{1}{2} W(y_1|u_1 \oplus u_2) W(y_2|u_2). \end{aligned}$$

Now will generalize these formulas.

Derivation of (4.87)
The probability expression

$$Prob(y_1^{2N}, u_1^{2i-2}|u_{2i-1})$$

can be written as

$$Prob(y_1^{2N}, u_1^{2i-2}|u_{2i-1}) = \frac{Prob(y_1^{2N}, u_1^{2i-2}, u_{2i-1})}{Prob(u_{2i-1})} \rightarrow$$

$$Prob(y_1^{2N}, u_1^{2i-2}|u_{2i-1}) = \frac{Prob(y_1^{2N}, u_1^{2i-1})}{Prob(u_{2i-1})}$$

in which using the property

$$p(x) = \sum_y p(x, y)$$

we get

$$Prob(y_1^{2N}, u_1^{2i-2}|u_{2i-1}) = \frac{1}{Prob(u_{2i-1})} \sum_{u_{2i}^{2N}} Prob(y_1^{2N}, u_1^{2N}) \tag{4.90}$$

which can be written as

$$Prob(y_1^{2N}, u_1^{2i-2}|u_{2i-1}) = \frac{1}{Prob(u_{2i-1})} \sum_{u_{2i}^{2N}} Prob(y_1^{2N}|u_1^{2N})Prob(u_1^{2N}) \tag{4.91}$$

where using $Prob(u_j) = 1/2$ and

$$Prob(u_1^{2N}) = Prob(u_1)Prob(u_2)\ldots Prob(u_{2N})$$

we get

$$Prob(u_1^{2N}) = \frac{1}{2^N}$$

then (4.91) happens to be

$$Prob(y_1^{2N}, u_1^{2i-2}|u_{2i-1}) = \frac{1}{2^{2N-1}} \sum_{u_{2i}^{2N}} Prob(y_1^{2N}|u_1^{2N}) \tag{4.92}$$

which can be written as

$$W_{2N}^{2i-1}(y_1^{2N}, u_1^{2i-2}|u_{2i-1}) = \frac{1}{2^{2N-1}} \sum_{u_{2i}^{2N}} W_{2N}(y_1^{2N}|u_1^{2N}) \tag{4.93}$$

where using the property

$$W_{2N}(\cdot) = W_N(\cdot)W_N(\cdot)$$

and referring to Fig. 4.7, we obtain

$$W_{2N}^{2i-1}(y_1^{2N}, u_1^{2i-2}|u_{2i-1}) = \frac{1}{2^{2N-1}} \sum_{u_{2i}^{2N}} W_N(y_1^N|s_{1,o}^N)W_N(y_{N+1}^{2N}|s_{1,e}^N) \tag{4.94}$$

leading to

$$W_{2N}^{2i-1}(y_1^{2N}, u_1^{2i-2}|u_{2i-1}) = \frac{1}{2^{2N-1}} \sum_{u_{2i}^{2N}} W_N(y_1^N|u_{1,o}^{2N} \oplus u_{1,e}^{2N})W_N(y_{N+1}^{2N}|u_{1,e}^{2N}) \tag{4.95}$$

which can be decomposed as

$$W_{2N}^{2i-1}(y_1^{2N}, u_1^{2i-2}|u_{2i-1}) = \frac{1}{2} \sum_{u_{2i,e}^{2N}} \frac{1}{2^{N-1}} W_N(y_{N+1}^{2N}|u_{1,e}^{2N}) \sum_{u_{2i,o}^{2N}} \frac{1}{2^{N-1}} W_N(y_1^N|u_{1,o}^{2N} \oplus u_{1,e}^{2N})$$

leading to

$$W_{2N}^{2i-1}(y_1^{2N}, u_1^{2i-2}|u_{2i-1}) = \sum_{u_{2i}} \frac{1}{2} \sum_{u_{2i+1,e}^{2N}} \frac{1}{2^{N-1}} W_N(y_{N+1}^{2N}|u_{1,e}^{2N}) \sum_{u_{2i+1,o}^{2N}} \frac{1}{2^{N-1}} W_N(y_1^N|u_{1,o}^{2N} \oplus u_{1,e}^{2N})$$

in which using the definition

$$W_N^i(y_1^N, u_1^{i-1}|u_i) = \frac{1}{2^{N-1}} \sum_{u_{i+1}^N} W_N(y_1^N|u_1^N)$$

we obtain

$$W_{2N}^{2i-1}(y_1^{2N}, u_1^{2i-2}|u_{2i-1}) = \sum_{u_{2i}} \frac{1}{2} W_N^i(y_1^N, u_{1,o}^{2i-2}$$
$$\oplus\, u_{1,e}^{2i-2}|u_{2i-1} \oplus u_{2i})W_N^i(y_{N+1}^{2N}, u_{1,e}^{2i-2}|u_{2i}) \tag{4.96}$$

and similarly for the channel W_{2N}^{2i}, considering the channel output, the conditional probability of the channel input u_{2i-1} can be recursively calculated as

$$Prob(y_1^{2N}, u_1^{2i-1}|u_{2i}) = \frac{1}{2} Prob(y_1^N, u_{1,o}^{2i-2} \oplus u_{1,e}^{2i-2}|u_{2i-1} \oplus u_{2i})Prob(y_{N+1}^{2N}, u_{1,e}^{2i-2}|u_{2i})$$

which can be written as

$$W_{2N}^{2i}(y_1^{2N}, u_1^{2i-1}|u_{2i}) = \frac{1}{2} W_N^i(y_1^N, u_{1,o}^{2i-2} \oplus u_{1,e}^{2i-2}|u_{2i-1} \oplus u_{2i}) W_N^i(y_{N+1}^{2N}, u_{1,e}^{2i-2}|u_{2i}). \tag{4.97}$$

Derivation of (4.89)
The probability expression

$$Prob(y_1^{2N}, u_1^{2i-1}|u_{2i})$$

can be written as

$$Prob(y_1^{2N}, u_1^{2i-1}|u_{2i}) = \frac{Prob(y_1^{2N}, u_1^{2i-1}, u_{2i})}{Prob(u_{2i})}$$

which can be expressed as

$$Prob(y_1^{2N}, u_1^{2i-1}|u_{2i}) = \frac{1}{Prob(u_{2i})} \sum_{u_{2i+1}^N} Prob(y_1^{2N}, u_1^{2N}) \rightarrow$$

$$Prob(y_1^{2N}, u_1^{2i-1}|u_{2i}) = \frac{1}{Prob(u_{2i})} \sum_{u_{2i+1}^N} Prob(y_1^{2N}|u_1^{2N}) Prob(u_1^{2N}) \tag{4.98}$$

where using $Prob(u_j) = 1/2$ and

$$Prob(u_1^{2N}) = Prob(u_1) Prob(u_2) \ldots Prob(u_{2N})$$

we get

$$Prob(u_1^{2N}) = \frac{1}{2^{2N}}.$$

Then, (4.98) happens to be

$$Prob(y_1^{2N}, u_1^{2i-1}|u_{2i}) = \frac{1}{2^{2N-1}} \sum_{u_{2i+1}^N} Prob(y_1^{2N}|u_1^{2N})$$

which can also be written as

$$W_{2N}^{2i}(y_1^{2N}, u_1^{2i-1}|u_{2i}) = \frac{1}{2^{2N-1}} \sum_{u_{2i+1}^N} W_{2N}(y_1^{2N}|u_1^{2N}) \tag{4.99}$$

where using the property

$$W_{2N}(\cdot) = W_N(\cdot) W_N(\cdot)$$

and referring to Fig. 4.7, we obtain

$$W_{2N}^{2i}(y_1^{2N}, u_1^{2i-1}|u_{2i}) = \frac{1}{2^{2N-1}} \sum_{u_{2i+1}^{2N}} W_N(y_1^N|s_{1,o}^N) W_N(y_{N+1}^{2N}|s_{1,e}^N)$$

which can be written as

$$W_{2N}^{2i}(y_1^{2N}, u_1^{2i-1}|u_{2i}) = \frac{1}{2} \sum_{u_{2i+1,o}^{2N}} \frac{1}{2^{N-1}} W_N(y_1^N|u_{1,o}^{2N} \oplus u_{1,e}^{2N}) \sum_{u_{2i+1,e}^{2N}} \frac{1}{2^{N-1}} W_N(y_{N+1}^{2N}|u_{1,e}^{2N})$$

in which using the definition

$$W_N^i(y_1^N, u_1^{i-1}|u_i) = \frac{1}{2^{N-1}} \sum_{u_{i+1}^N} W_N(y_1^N|u_1^N)$$

we obtain

$$W_{2N}^{2i}(y_1^{2N}, u_1^{2i-1}|u_{2i}) = \frac{1}{2} W_N^i(y_1^N, u_{1,o}^{2i-2} \oplus u_{1,e}^{2i-2}|u_{2i-1} \oplus u_{2i}) W_N^i(y_{N+1}^{2N}, u_{1,e}^{2i-2}|u_{2i}). \tag{4.100}$$

4.6 Recursive Calculation of Likelihood Ratio

In this section, we will derive the recursive expression for the likelihood ratio calculation of the split channel input.

The likelihood ratio for the decoding of ith information bit is defined as

$$L_N^i(y_1^N, u_1^{i-1}) \triangleq \frac{Prob(y_1^N, u_1^{i-1}|u_i = 0)}{Prob(y_1^N, u_1^{i-1}|u_i = 1)}$$

which can also be written as

$$L_N^i(y_1^N, u_1^{i-1}) \triangleq \frac{W_N^i(y_1^N, u_1^{i-1}|u_i = 0)}{W_N^i(y_1^N, u_1^{i-1}|u_i = 1)}. \tag{4.101}$$

Proposition 4.2 *The likelihood ratio for the decoding of ith information bit can be recursively computed using*

$$L_N^{2i-1}(y_1^N, u_1^{2i-2}) = \frac{L_{N/2}^i(y_1^{N/2}, u_{1,o}^{2i-2} \oplus u_{1,e}^{2i-2}) L_{N/2}^i(y_{N/2+1}^N, u_{1,e}^{2i-2}) + 1}{L_{N/2}^i(y_1^{N/2}, u_{1,o}^{2i-2} \oplus u_{1,e}^{2i-2}) + L_{N/2}^i(y_{N/2+1}^N, u_{1,e}^{2i-2})} \tag{4.102}$$

and

$$L_N^{2i}(y_1^N, u_1^{2i-1}) = \left[L_{N/2}^i(y_1^{N/2}, u_{1,o}^{2i-2} \oplus u_{1,e}^{2i-2})\right]^{1-2u_{2i-1}} L_{N/2}^i(y_{N/2+1}^N, u_{1,e}^{2i-2}). \tag{4.103}$$

Proof 4.2 Let's first show (4.102). Using (4.101), we can write the left hand side of (4.102) as

$$L_N^{2i-1}(y_1^N, u_1^{2i-2}) = \frac{W_N^{2i-1}(y_1^N, u_1^{2i-2}|u_{2i-1}=0)}{W_N^{2i-1}(y_1^N, u_1^{2i-2}|u_{2i-1}=1)} \tag{4.104}$$

in which employing (4.96), we get

$$L_N^{2i-1}(y_1^N, u_1^{2i-2}) = \frac{\sum_{u_{2i}} \frac{1}{2} W_N^i(y_1^N, u_{1,o}^{2i-2} \oplus u_{1,e}^{2i-2}|0 \oplus u_{2i}) W_N^i(y_{N+1}^{2N}, u_{1,e}^{2i-2}|u_{2i})}{\sum_{u_{2i}} \frac{1}{2} W_N^i(y_1^N, u_{1,o}^{2i-2} \oplus u_{1,e}^{2i-2}|1 \oplus u_{2i}) W_N^i(y_{N+1}^{2N}, u_{1,e}^{2i-2}|u_{2i})}. \tag{4.105}$$

Let

$$A = (y_1^N, u_{1,o}^{2i-2} \oplus u_{1,e}^{2i-2}) B = y_{N+1}^{2N}, u_{1,e}^{2i-2} \tag{4.106}$$

then (4.105) can be written as

$$L_N^{2i-1}(y_1^N, u_1^{2i-2}) = \frac{\frac{1}{2} W_N^i(A|0 \oplus 0) W_N^i(B|0) + \frac{1}{2} W_N^i(A|0 \oplus 1) W_N^i(B|1)}{\frac{1}{2} W_N^i(A|1 \oplus 0) W_N^i(B|0) + \frac{1}{2} W_N^i(A|1 \oplus 0) W_N^i(B|0)} \tag{4.107}$$

which can be simplified as

$$L_N^{2i-1}(y_1^N, u_1^{2i-2}) = \frac{W_N^i(A|0) W_N^i(B|0) + W_N^i(A|1) W_N^i(B|1)}{W_N^i(A|1) W_N^i(B|0) + W_N^i(A|0) W_N^i(B|1)}. \tag{4.108}$$

Dividing the numerator and denominator of (4.108) by

$$W_N^i(A|1) W_N^i(B|1) \tag{4.109}$$

we get

$$L_N^{2i-1}(y_1^N, u_1^{2i-2}) = \frac{\frac{W_N^i(A|0) W_N^i(B|0)}{W_N^i(A|1) W_N^i(B|1)} + 1}{\frac{W_N^i(A|1) W_N^i(B|0)}{W_N^i(A|1) W_N^i(B|1)} + \frac{W_N^i(A|0) W_N^i(B|1)}{W_N^i(A|1) W_N^i(B|1)}}$$

which is simplified as

$$L_N^{2i-1}(y_1^N, u_1^{2i-2}) = \frac{\frac{W_N^i(A|0)}{W_N^i(A|1)} \times \frac{W_N^i(B|0)}{W_N^i(B|1)} + 1}{\frac{W_N^i(B|0)}{W_N^i(B|1)} + \frac{W_N^i(A|0)}{W_N^i(A|1)}} \tag{4.110}$$

in which substituting the values of A, B and using the likelihood definition in (4.101), we obtain

$$L_N^{2i-1}(y_1^N, u_1^{2i-2}) = \frac{L_{N/2}^i(y_1^{N/2}, u_{1,o}^{2i-2} \oplus u_{1,e}^{2i-2})L_{N/2}^i(y_{N/2+1}^N, u_{1,e}^{2i-2}) + 1}{L_{N/2}^i(y_1^{N/2}, u_{1,o}^{2i-2} \oplus u_{1,e}^{2i-2}) + L_{N/2}^i(y_{N/2+1}^N, u_{1,e}^{2i-2})}. \quad (4.111)$$

Let's now show (4.103) Using (4.101) we can write the left hand side of (4.103) as

$$L_N^{2i}(y_1^N, u_1^{2i-2}) = \frac{W_N^{2i}(y_1^N, u_1^{2i-1}|u_{2i} = 0)}{W_N^{2i}(y_1^N, u_1^{2i-1}|u_{2i} = 1)} \quad (4.112)$$

in which using (4.100), we get

$$L_N^{2i}(y_1^N, u_1^{2i-2}) = \frac{W_N^i(y_1^{N/2}, u_{1,o}^{2i-2} \oplus u_{1,e}^{2i-2}|u_{2i-1} \oplus 0)W_N^i(y_{N/2+1}^N, u_{1,e}^{2i-2}|0)}{W_N^i(y_1^{N/2}, u_{1,o}^{2i-2} \oplus u_{1,e}^{2i-2}|u_{2i-1} \oplus 1)W_N^i(y_{N/2+1}^N, u_{1,e}^{2i-2}|1)}. \quad (4.113)$$

For $u_{2i-1} = 0$, (4.113) happens to be

$$L_N^{2i}(y_1^N, u_1^{2i-2}) = \frac{W_N^i(y_1^{N/2}, u_{1,o}^{2i-2} \oplus u_{1,e}^{2i-2}|0)W_N^i(y_{N/2+1}^N, u_{1,e}^{2i-2}|0)}{W_N^i(y_1^{N/2}, u_{1,o}^{2i-2} \oplus u_{1,e}^{2i-2}|1)W_N^i(y_{N/2+1}^N, u_{1,e}^{2i-2}|1)}$$

which can be written as

$$L_N^{2i}(y_1^N, u_1^{2i-2}) = L_{N/2}^i(y_1^{N/2}, u_{1,o}^{2i-2} \oplus u_{1,e}^{2i-2})L_{N/2}^i(y_{N/2+1}^N, u_{1,e}^{2i-2}) \quad (4.114)$$

and for $u_{2i-1} = 1$, (4.113) happens to be

$$L_N^{2i}(y_1^N, u_1^{2i-2}) = \frac{W_N^i(y_1^{N/2}, u_{1,o}^{2i-2} \oplus u_{1,e}^{2i-2}|1)W_N^i(y_{N/2+1}^N, u_{1,e}^{2i-2}|0)}{W_N^i(y_1^{N/2}, u_{1,o}^{2i-2} \oplus u_{1,e}^{2i-2}|0)W_N^i(y_{N/2+1}^N, u_{1,e}^{2i-2}|1)}$$

which can be written as

$$L_N^{2i}(y_1^N, u_1^{2i-2}) = \left[L_{N/2}^i(y_1^{N/2}, u_{1,o}^{2i-2} \oplus u_{1,e}^{2i-2})\right]^{-1} L_{N/2}^i(y_{N/2+1}^N, u_{1,e}^{2i-2}). \quad (4.115)$$

We can combine (4.114) and (4.115) into a single expression as

$$L_N^{2i}(y_1^N, u_1^{2i-2}) = \left[L_{N/2}^i(y_1^{N/2}, u_{1,o}^{2i-2} \oplus u_{1,e}^{2i-2})\right]^{1-2u_{2i-1}} L_{N/2}^i(y_{N/2+1}^N, u_{1,e}^{2i-2}). \quad (4.116)$$

4.6.1 Bit Value Decision

Once we find the conditional probabilities for information bits, we can decide on the value of the information bits as

$$u_i = \begin{cases} 0 & \text{if } L_N^i(y_1^N, u_1^{i-1}) \geq 1 \\ 1 & \text{otherwise} \end{cases} \tag{4.117}$$

where $L_N^i(y_1^N, u_1^{i-1})$ can be calculated recursively as in (4.102) and (4.103).

Problems

(1) Show the split channels W_8^1, W_8^3 and W_8^5 by bold lines in Fig. P4.1.
(2) The polar encoder using the alternative construction method is depicted in Fig. P4.2. Show the split channels W_4^1, W_4^2 and W_4^4 by bold line using Fig. P4.2.

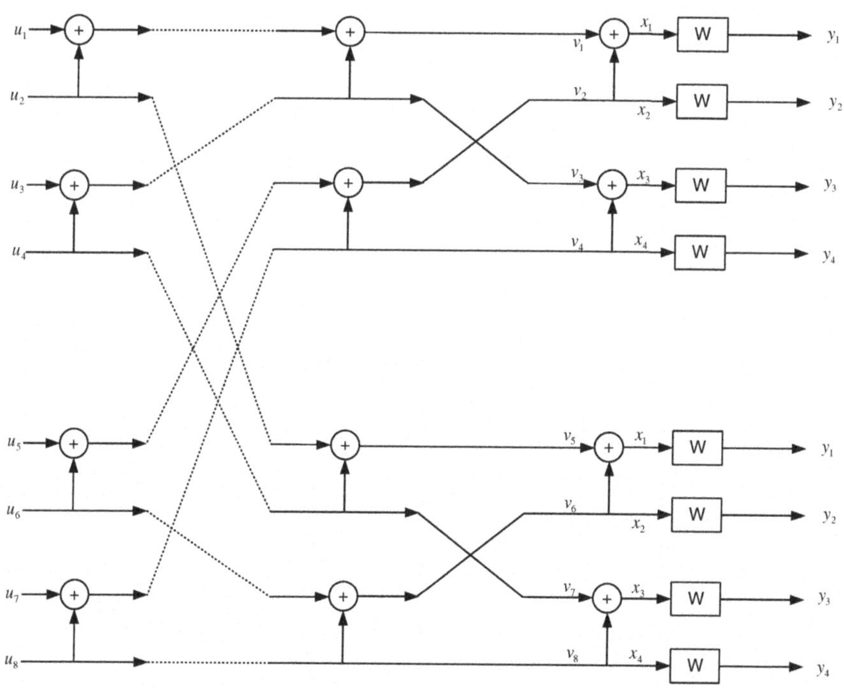

Fig. P4.1 Polar encoder structure for P1

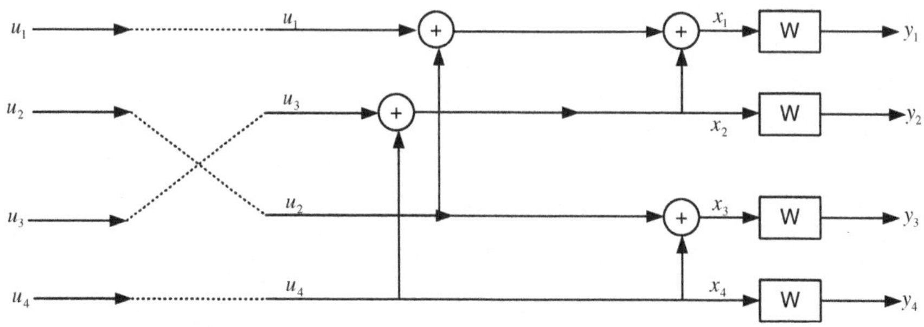

Fig. P4.2 Polar encoder structure for P2

Chapter 5
Polarization Rate and Performance of Polar Codes

In this chapter, we will mathematically state the channel polarization theorem and prove it. In addition, the performance of polar codes will be mathematically analyzed.

The proof of the channel polarization theorem is mathematically highly involved concept. The reader should have good knowledge of probability and stochastic process. Before giving the theorem and its proof, we will review some background information so that we can understand the proof.

5.1 σ-Field or σ-Algebra

An experiment can be classified as discrete or continuous. Flipping a coin can be considered as a discrete experiment, on the other hand, measuring the temperature of weather can be accepted as a continuous experiment. For the simplicity of explanation, we will consider the discrete experiments.

A discrete experiment has a number of countable outcomes. The set consisting of distinct outputs of a discrete experiment is called sample space.

Example 5.1 For the throwing a die experiment, the sample space is $S = \{f_1, f_2, f_2, f_4, f_5, f_6\}$, and for the flipping a coin experiment the sample space is $S = \{h, t\}$.

Example 5.2 Let $S = \{s_1, s_2, s_3\}$ be the sample space of an experiment, then the number of subsets of S equals to $2^3 = 8$, and the subsets of S can be written as

$$\phi, \{s_1\}, \{s_2\}, \{s_3\}, \{s_1, s_2\}, \{s_1, s_3\}, \{s_2, s_3\}, S \tag{5.1}$$

where ϕ is the empty set.

O. Gazi, *Polar Codes*, Springer Topics in Signal Processing 15,
https://doi.org/10.1007/978-981-13-0737-9_5

σ-field or *σ-algebra*:

Consider a collection of subsets of S, and let's denote the collection of subsets of S by F. The set F is called a *σ-field or σ-algebra,* if it satisfies the following conditions:

(1) $\phi, S \in F$, i.e., empty set and sample space should be included in F.
(2) If $A \in F$, then we should have $A^c \in F$, i.e., if a subset is an element of F, then its complement A^c should also be an element of F.
(3) If $A_1, A_2, \ldots \in F$, then we should have $\bigcup_i A_i \in F$.

F can also be called event space, and the elements of F are called events.

Example 5.3 Let S be the sample space of an experiment, then we can define the following *σ-fields.*

(1) $F = \{\phi, S\}$ is the simplest *σ-field* we can form.
(2) We can define F considering all the possible subsets of S, and F contains $2^{|S|}$ elements, where $|S|$ indicated the number of elements in S. For instance, if $S = \{s_1, s_2, s_3, s_4\}$, then $|S| = 4$, and the *σ-field* F contains $2^4 = 16$ subsets, i.e., elements.
(3) If A is a subset of S, i.e., $A \subset S$, then we can form the *σ-field* including A as $F = \{\phi, A, A^c, S\}$.

If we have a *σ-field* $F = \{\phi, A, A^c, S\}$, we can consider this field to be generated by A and denote it by $\sigma(A)$, i.e., $\sigma(A) = \{\phi, A, A^c, S\}$.

Example 5.4 Let $S = \{s_1, s_2, s_3, s_4, s_5, s_6\}$ be the sample space of a discrete experiment. Considering the subset $A = \{s_1, s_3, s_5\}$, we can form the a *σ-field*, i.e., event space, as

$$F = \{\phi, A, A^c, S\} \rightarrow F = \sigma\{A\} = \{\phi, \{s_1, s_3, s_5\}, \{s_2, s_4, s_6\}, \{s_1, s_2, s_3, s_4, s_5, s_6\}\}.$$

5.1.1 Borel Field or Borel σ-Field

For a continuous experiment, the sample space includes uncountable number of elements. For this reason, we indicate the sample spaces by real number intervals.

Example 5.5 The sample space for the weather temperature in the world can be written as

$$S = [-20\,60]$$

which is a real number interval containing uncountable number of elements.

Borel Field or Borel σ-field:

Let $S = [a\, b]$ be the sample space, the Borel field, i.e., $\mathcal{B}$, over $[a\, b]$ is the smallest σ-field containing all the closed and open intervals in $[a\, b]$. Each element of the Borel field is called a Borel set. If the Borel field is generated by a number of K intervals, then the generated Borel field is indicated by $\mathcal{B} = \sigma(K)$.

5.1.2 Measure Function and Measure Space

A measure function $g(\cdot)$, which is a nonnegative function, is defined on a *σ-field,* i.e., on F, satisfying the following properties

(1) If $A \in F$, then $0 \leq g(A) < \infty$.
(2) $g(\phi) = 0$.
(3) If $A, B \in F$ such that $A \cap B = \phi$, then $g(A \cap B) = g(A) + g(B)$

A measure space is a *σ-field* with a measure function defined on the elements of *σ-field.*

Example 5.6 A Borel sigma algebra with length measure can be considered as a measure space.

5.1.3 Probability Space

Let S be the sample space of an experiment, and F is the *σ-field*, i.e., event space, containing all the subsets of S, and $Prob(\cdot)$ is a measure function defined on F satisfying

(1) If $A \in F$, then $0 \leq Prob(A) \leq 1$.
(2) $Prob(\phi) = 0$, $Prob(S) = 1$.
(3) If $A, B \in F$ such that $A \cap B = \phi$, then we have $Prob(A \cup B) = Prob(A) + Prob(B)$.

The triple $(S, F, Prob)$ is named as a probability space. For the simplicity of the notation, we can use P to indicate the function $Prob$.

5.2 Random Variable

A random variable is nothing but a real valued function whose inputs are the elements of the sample space of an experiment.

Assume that we have a three sided die. Our sample space can be written as

$$S = \{f_1, f_2, f_3\} \tag{5.2}$$

where f_1, f_2, f_3 are simple outputs of the experiment, and for a fair die, simple outputs have equal probability of occurrence. For this example, we have

$$Prob(f_1) = Prob(f_2) = Prob(f_3) = \frac{1}{3}. \tag{5.3}$$

We can define a random variable $\tilde{X}(\cdot)$ on S as

$$\tilde{X}(f_1) = -2 \quad \tilde{X}(f_2) = 1 \quad \tilde{X}(f_3) = 2.5. \tag{5.4}$$

The range set or value set of the $\tilde{X}$ random variable can be indicated as

$$R_{\tilde{X}} = \{-2, 1, 2.5\}. \tag{5.5}$$

The expression

$$\left\{s_i | \tilde{X}(s_i) = x\right\} \tag{5.6}$$

denotes a subset of S consisting of those s_i which satisfies $\tilde{X}(s_i) = x$. For simplicity of notation, an equivalent expression of

$$\left\{s_i | \tilde{X}(s_i) = x\right\} \tag{5.7}$$

can be given as

$$\tilde{X} = x. \tag{5.8}$$

The probability mass function of discrete random variable $\tilde{X}$ is defined as

$$p(x) = Prob\left(\tilde{X} = x\right) \tag{5.9}$$

where x is a value in the range set $R_{\tilde{X}}$.

Example 5.7 The sample space of an experiment is given as $S = \{s_1, s_2, s_3, s_4\}$. The simple outcomes of S have equal probabilities. The random variable $\tilde{X}(\cdot)$ is defined on S as

$$\tilde{X}(s_1) = -1 \quad \tilde{X}(s_2) = 1 \quad \tilde{X}(s_3) = 1 \quad \tilde{X}(s_4) = 2.$$

Write the range set of $\tilde{X}$ and calculate the probability mass function of $\tilde{X}$.

Solution 5.7 The range set can be written as $R_{\tilde{X}} = \{-1, 1, 2\}$. Using the definition of probability mass function

$$p(x) = Prob\left(\tilde{X} = x\right)$$

we can calculate the probability mass function as

$$p(x = -1) = Prob\left(\tilde{X} = -1\right) \rightarrow p(x = -1) = Prob(s_1) \rightarrow p(x = -1) = \frac{1}{4}$$
$$p(x = 1) = Prob\left(\tilde{X} = 1\right) \rightarrow p(x = 1) = Prob(s_2, s_3) \rightarrow p(x = 1) = \frac{2}{4}$$
$$p(x = 2) = Prob\left(\tilde{X} = 2\right) \rightarrow p(x = 2) = Prob(s_4) \rightarrow p(x = 2) = \frac{1}{4}.$$

Mean and Variance of a Random Variable

The probabilistic mean or expected value of a discrete random variable is calculated as

$$E\left(\tilde{X}\right) = \sum_x x\, p(x) \tag{5.10}$$

which is usually denoted by m, i.e., $m = E(\tilde{X})$.

The variance of a discrete random variable is calculated as

$$Var\left(\tilde{X}\right) = E\left(\tilde{X}^2\right) - m^2 \tag{5.11}$$

where $E(\tilde{X}^2)$ is calculated as

$$E\left(\tilde{X}^2\right) = \sum_x x^2 p(x). \tag{5.12}$$

Bernoulli Random Variable:

The Bernoulli Random variable $\tilde{X}$ has the range set $R_{\tilde{X}} = \{0, 1\}$, and the probability mass function of the Bernoulli random variable is given as

$$p(x) = \begin{cases} p & x = 1 \\ 1 - p & x = 0 \\ 0 & otherwise. \end{cases} \tag{5.13}$$

5.3 Random or Stochastic Process

Let $S = \{s_1, s_2, \ldots, s_M\}$ be the sample space of a discrete experiment. A random variable $\tilde{X}(s_i)$ is a real valued function whose inputs are simple events s_i, i.e., the function $\tilde{X}(s_i)$ maps the simple events s_i to real numbers.

Example 5.8 For the sample space $S = \{s_1, s_2, s_3\}$, the random variable $\tilde{X}(s_i)$ is defined as

$$\tilde{X}(s_1) = -1 \quad \tilde{X}(s_2) = 1 \quad \tilde{X}(s_3) = -1$$

A stochastic process $\tilde{X}(s_i, t)$, on the other hand, maps the simple events of a sample space to time varying functions, i.e.,

$$\tilde{X}(s_1, t) = g_1(t) \quad \tilde{X}(s_2, t) = g_2(t) \ldots \tilde{X}(s_M, t) = g_M(t).$$

Example 5.9 For the sample space $S = \{s_1, s_2, s_3\}$, the random process $\tilde{X}(s_i, t)$ is defined as

$$\tilde{X}(s_1, t) = \sin t \quad \tilde{X}(s_2, t) = e^{-t} \cos t \quad \tilde{X}(s_3, t) = \sin 2t$$

For the simplicity of notation $\tilde{X}$ will be used for $\tilde{X}(s_i)$, and $\tilde{X}(t)$ will be used for $\tilde{X}(s_i, t)$.

The random process $\tilde{X}(t)$ indicates a different random variable for different t values, i.e., $\tilde{X}(t_1)$, $\tilde{X}(t_2)$, ... are different random variables, and these random variables are defined on the same sample space.

The mean value of a random process is also a function of time, and is calculated as

$$m(t) = E\left(\tilde{X}(t)\right) = \sum_{x_t} x_t p_t(x_t) \tag{5.14}$$

Example 5.10 For the sample space $S = \{s_1, s_2, s_3\}$, the random process $\tilde{X}(s_i, t)$ is defined as

$$\tilde{X}(s_1, t) = g_1(t) \; \tilde{X}(s_2, t) = g_2(t) \quad \tilde{X}(s_3, t) = g_3(t)$$

and the graphs of the functions $g_1(t)$, $g_2(t)$, $g_3(t)$ are depicted in Fig. 5.1.

At time instant t_1, we have

$$\tilde{X}(s_1, t_1) = 0.7 \; \tilde{X}(s_2, t_1) = 0.6 \; \tilde{X}(s_3, t_1) = 0.5$$

Fig. 5.1 The graphs of the functions $g_1(t)$, $g_2(t)$, and $g_3(t)$

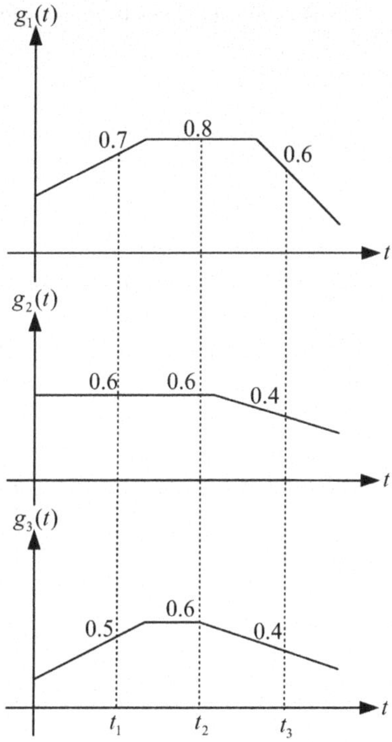

The range set of the random variable $\tilde{X}(t_1)$ can be written as

$$R_{t_1} = \{0.7, 0.6, 0.5\}.$$

We can find the probability mass function of the random variable $\tilde{X}(t_1)$ using the definition

$$p_1(x) = Prob\left(\tilde{X}(t_1) = x\right) \tag{5.15}$$

as

$$p_1(x = 0.7) = Prob\left(\tilde{X}(t_1) = 0.7\right) \rightarrow p_1(x = 0.7) = Prob(s_1) \rightarrow p_1(x = 0.7) = \frac{1}{3}$$
$$p_1(x = 0.6) = Prob\left(\tilde{X}(t_1) = 0.6\right) \rightarrow p_1(x = 0.6) = Prob(s_2) \rightarrow p_1(x = 0.6) = \frac{1}{3}$$
$$p_1(x = 0.5) = Prob\left(\tilde{X}(t_1) = 0.5\right) \rightarrow p_1(x = 0.5) = Prob(s_3) \rightarrow p_1(x = 0.5) = \frac{1}{3}.$$

The range set of the random variable $\tilde{X}(t_2)$ can be written as

$$R_{t_2} = \{0.8, 0.6\}.$$

The probability mass function of the random variable $\tilde{X}(t_2)$ from the definition

$$p_2(x) = Prob\left(\tilde{X}(t_2) = x\right) \tag{5.16}$$

can be found as

$$p_2(x = 0.8) = Prob\left(\tilde{X}(t_2) = 0.8\right) \rightarrow p_2(x = 0.8) = Prob(s_1) \rightarrow p_2(x = 0.8) = \frac{1}{3}$$

$$p_2(x = 0.6) = Prob\left(\tilde{X}(t_1) = 0.6\right) \rightarrow p_2(x = 0.6) = Prob(s_2, s_3) \rightarrow p_2(x = 0.6) = \frac{2}{3}$$

Finally, the range set of the random variable $\tilde{X}(t_3)$ can be written as

$$R_{t_3} = \{0.6, 0.4\}.$$

The probability mass function of the random variable $\tilde{X}(t_3)$ from the definition

$$p_3(x) = Prob\left(\tilde{X}(t_3) = x\right) \tag{5.17}$$

can be found as

$$p_3(x = 0.6) = Prob\left(\tilde{X}(t_3) = 0.6\right) \rightarrow p_3(x = 0.6) = Prob(s_1) \rightarrow p_3(x = 0.6) = \frac{1}{3}$$

$$p_3(x = 0.4) = Prob\left(\tilde{X}(t_3) = 0.4\right) \rightarrow p_3(x = 0.4) = Prob(s_2, s_3) \rightarrow p_3(x = 0.4) = \frac{2}{3}$$

To sum it up, the random process $\tilde{X}(t)$ at time instant t_1, t_2, t_3 generates the random variables $\tilde{X}(t_1)$, $\tilde{X}(t_2)$, $\tilde{X}(t_3)$ whose range sets and probability mass functions are found as

$$\begin{array}{lll} R_{t_1} = \{0.7, 0.6, 0.5\} & R_{t_2} = \{0.8, 0.6\} & R_{t_3} = \{0.6, 0.4\} \\ p_1 = \left\{\frac{1}{3}, \frac{1}{3}, \frac{1}{3}\right\} & p_2 = \left\{\frac{1}{3}, \frac{2}{3}\right\} & p_3 = \left\{\frac{1}{3}, \frac{2}{3}\right\}. \end{array} \tag{5.18}$$

Using the found values in (5.18), we can calculate the expected values of the random variables as

$$m(t_1) = E\left(\tilde{X}(t_1)\right) = \sum_{x_{t_1}} x_{t_1} p_{t_1}(x_{t_1}) \rightarrow m(t_1) = 0.7 \times \frac{1}{3} + 0.6 \times \frac{1}{3} + 0.5 \times \frac{1}{3} \rightarrow m(t_1) = 0.6$$

$$m(t_2) = E\left(\tilde{X}(t_2)\right) = \sum_{x_{t_2}} x_{t_2} p_{t_2}(x_{t_2}) \rightarrow m(t_2) = 0.8 \times \frac{1}{3} + 0.6 \times \frac{2}{3} \rightarrow m(t_2) \approx 0.67$$

$$m(t_1) = E\left(\tilde{X}(t_1)\right) = \sum_{x_{t_3}} x_{t_3} p_{t_3}(x_{t_3}) \rightarrow m(t_3) = 0.6 \times \frac{1}{3} + 0.4 \times \frac{2}{3} \rightarrow m(t_3) \approx 0.47.$$

Thus, we see that expected value of random process is a time dependent quantity.

The random process $\tilde{X}(t)$ can also be indicated as $\tilde{X}_t$, or using a different parameter, it can be indicated as $\tilde{X}_n$. If we consider the time instant $t_1, t_2, \ldots, t_N$, we have a sequence of random variables $\tilde{X}_1, \tilde{X}_2, \ldots, \tilde{X}_N$ and these random variables can be written as a vector of random variables

$$\tilde{\boldsymbol{X}} = \left[\tilde{X}_1 \tilde{X}_2, \ldots \tilde{X}_N\right]. \tag{5.19}$$

Example 5.11 Assume that there are N independent identically distributed, i.e., IID, Bernoulli random variables $\tilde{X}_1, \tilde{X}_2, \ldots, \tilde{X}_N$ which appear at the time instants $t_1, t_2, \ldots, t_N$. Then, the sequence of Bernoulli random variables can be considered as a stochastic process.

5.3.1 Cylinder Sets

Let $B = \{0, 1\}$ and

$$S = \{0, 1\}^N \rightarrow S = \{\bar{w} | \bar{w} = \{w_1, w_2, \ldots\} : w_i \in \{0, 1\}\} \tag{5.20}$$

i.e., S is the set of binary vectors of length N at most. Let S_c be the collection of all subsets of S on only finitely many co-ordinates. Such sets are called cylinders.

Alternatively, we can define a cylinder set as

$$S_c = \{\bar{w} | \bar{w} = \{w_{k1} = \epsilon_1, w_{k2} = \epsilon_2, \ldots w_{kn} = \epsilon_n\}\} \tag{5.21}$$

where

$$n \geq 1, k_1 < k_2 < \cdots < k_n \quad \text{and} \quad \epsilon_i \in \{0, 1\}. \tag{5.22}$$

Example 5.12 An example for a cylinder set can be given as

$$S_c = \{000, 001, 010, 100, 101, 110, 111\}.$$

5.3.2 Measurable Function

The function $\tilde{X}$ is measurable, if $A \in F$ and $\tilde{X}(A) = w$, then we have $A = \tilde{X}^{-1}(w)$. Random variables are measurable functions.

Example 5.13 The function defined by

$$\tilde{X}(A) = \begin{cases} 2 & \text{if } A = \{1, 2\} \\ 3 & \text{if } A = \{3, 4\} \end{cases} \tag{5.23}$$

is measurable considering the σ-*field* $F_1 = \{\phi, \{1, 2\}, \{3, 4\}, \{1, 2, 3, 4\}\}$ but it is not measurable considering $F_2 = \{\phi, \{1, 3\}, \{2, 4\}, \{1, 2, 3, 4\}\}$.

The mathematical expression $\sigma(\tilde{X})$ indicates the sub σ-*field* F_1, such that $F_1 \subset F$, and F_1 includes A i.e., $\sigma(\tilde{X}) = \{A|\tilde{X}(A) = w\}$.

Filtration

Given a probability space (S, F, P), an increasing family of sub σ-*fields* $\{F_n, n \geq 0\}$ satisfying the constraint

$$F_0 \subseteq F_1 \subseteq \cdots F \tag{5.24}$$

is named as filtration.

5.3.3 *Adopted Process*

The random process $\tilde{X}_n$ is adopted to the filtration F_n if $\tilde{X}_n$ is F_n measurable.

5.3.4 *Martingale Process*

The random process $\tilde{X}_n$ considering F_n is martingale if it satisfies the following:

(1) The random process $\tilde{X}_n$ is adopted.

(2)

$$E\left(\left|\tilde{X}_n\right|\right) \leq \infty. \tag{5.25}$$

(3)

$$E\left(\tilde{X}_n|F_{n-1}\right) = \tilde{X}_{n-1} \tag{5.26}$$

which implies that

$$E\left(\tilde{X}_n|\tilde{X}_{n-1}, \tilde{X}_{n-2,} \cdots, \tilde{X}_0\right) = \tilde{X}_{n-1}. \tag{5.27}$$

Example 5.14 Let $\{\tilde{X}_i, i \geq 1\}$ be a sequence of independent and identically distributed, i.e., iid, zero mean random variables, i.e., $E(\tilde{X}_i) = 0$. Shown that the sum of random variables $\tilde{X}_i$ is a martingale, i.e., if

$$\tilde{Y}_N = \sum_{i=1}^{N} \tilde{X}_i \quad \text{with} \quad \tilde{Y}_0 = 0$$

then

$$\left\{\tilde{Y}_N,\ N \geq 0\right\}$$

is a martingale.

Solution 5.13 Let $F_N = \sigma(\tilde{X}_1, \ldots, \tilde{X}_N)$ be the σ-field generated by the first N random variables $\tilde{X}_i$'s. We can calculate

$$E\left(\tilde{Y}_{N+1}|\tilde{X}, \ldots, \tilde{X}_N\right)$$

as in

$$\begin{aligned} E\left(\tilde{Y}_{N+1}|\tilde{X}, \ldots, \tilde{X}_N\right) &= E\left(\tilde{X}_{N+1} + \tilde{Y}_N|\tilde{X}_1, \ldots, \tilde{X}_N\right) \\ &= E\left(\tilde{X}_{N+1}|\tilde{X}_1, \ldots, \tilde{X}_N\right) + E\left(\tilde{Y}_N|\tilde{X}_1, \ldots, \tilde{X}_N\right) \\ &= E\left(\tilde{X}_{N+1}\right) + \tilde{Y}_N \\ &= \tilde{Y}_N. \end{aligned}$$

Thus,

$$E\left(\tilde{Y}_{N+1}|\tilde{X}, \ldots, \tilde{X}_N\right) = \tilde{Y}_N$$

which implies that $\tilde{Y}_N$ is a martingale process.

Exercise *Let $\{\tilde{X}_i,\ i \geq 1\}$ be a sequence of independent and identically distributed random variables, with $E(\tilde{X}_i) = 1$. Shown that the product of random variables $\tilde{X}_i$ is a martingale, i.e., if*

$$\tilde{S}_N = \prod_{i=1}^{N} \tilde{X}_i$$

then

$$\left\{\tilde{S}_N, N \geq 0\right\}$$

is a martingale.

5.3.5 Super-Martingale Process

The random process $\tilde{X}_n$ is a super-martingale if it satisfies

$$E\left(\tilde{X}_n | F_{n-1}\right) \leq \tilde{X}_{n-1} \; n \geq 1. \tag{5.28}$$

5.3.6 Sub-Martingale Process

The random process $\tilde{X}_n$ is a sub-martingale if it satisfies

$$E\left(\tilde{X}_n | F_{n-1}\right) \geq \tilde{X}_{n-1} \; n \geq 1. \tag{5.29}$$

The super-martingale decreases in time on average, on the other hand, a sub-martingale increases in time on average.

5.3.7 Martingale Process Convergence Theorem

If the random process $\tilde{X}_n$ is a martingale and $E\left(\left|\tilde{X}_n\right|\right) \leq K$, then

$$\lim_{n \to \infty} \tilde{X}_n \tag{5.30}$$

exits with probability 1, and it is finite.

5.4 Channel Polarization Theorem

In Chap. 1, recursive construction of polar encoder structures are explained in details, and in Chap. 4 we gave information about split channels, and we showed that when two split channels W_N^i are combined we obtain the split channels W_{2N}^{2i-1} and W_{2N}^{2i}, and this operation is expressed as

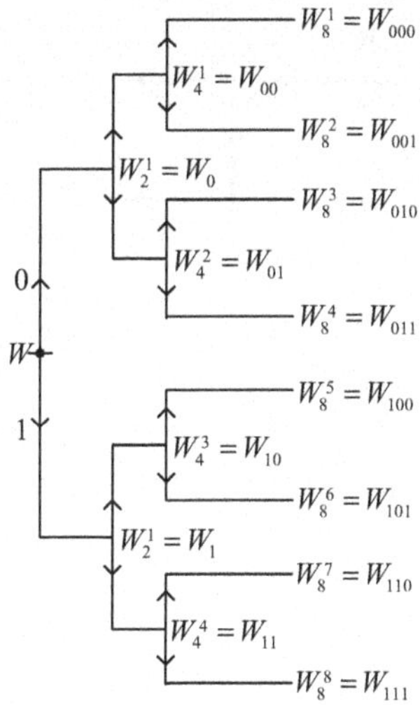

Fig. 5.2 The tree structure for recursive channel construction

$$\left(W_N^i, W_N^i\right) \rightarrow \left(W_{2N}^{2i-1}, W_{2N}^{2i}\right).$$

And in Fig. 4.23, the recursive channel formation is shown in butterfly structure which can be redrawn as a tree structure as in Fig. 5.2.

As it is indicated in Fig. 5.2 that we can denote the split channel

$$W_{2^n}^i \tag{5.31}$$

by

$$W_{b_1 b_2 \cdots b_n} \tag{5.32}$$

such that

$$i = 1 + \sum_{j=1}^{n} b_j 2^{n-j}. \tag{5.33}$$

For instance, we can indicate

$$W_8^5$$

by

$$W_{100}$$

since

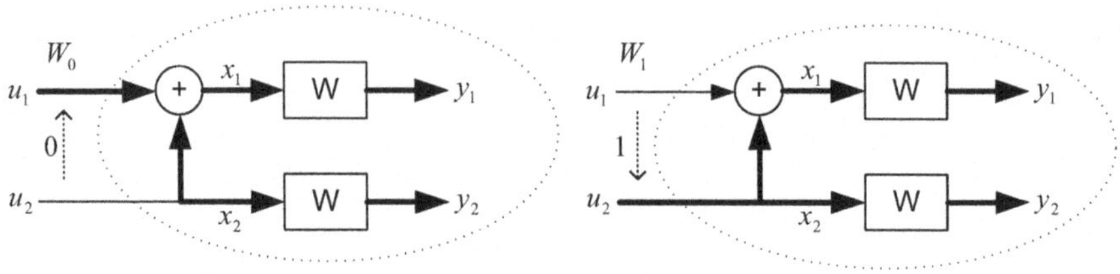

Fig. 5.3 The $W_{\bar{b}}$ representations of the split channels W_2^1 and W_2^2

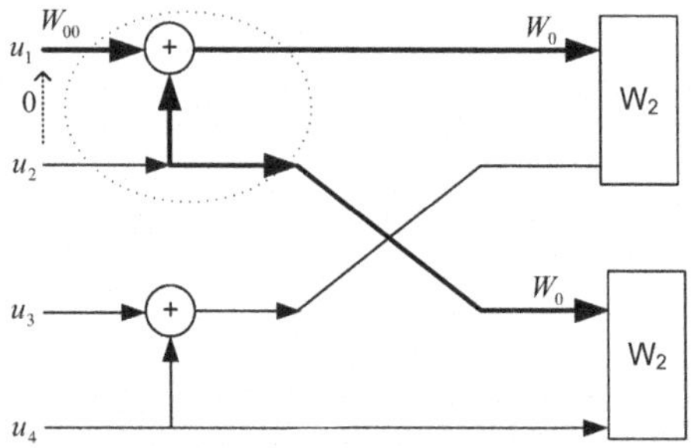

Fig. 5.4 The $W_{\bar{b}} = W_{00}$ representations of the split channel W_4^1

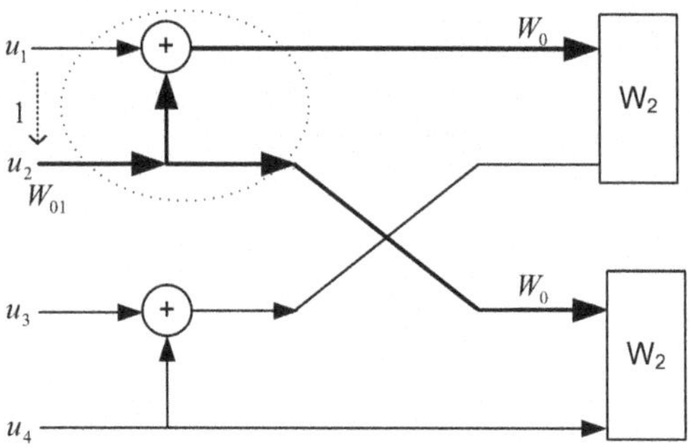

Fig. 5.5 The $W_{\bar{b}} = W_{01}$ representations of the split channel W_4^2

$$5 = 1 + \sum_{j=1}^{n} b_j 2^{n-j} \text{ where } n = 3 \text{ and } b_1 b_2 b_3 = 100. \tag{5.34}$$

We can also denote $W_{b_1 b_2 \cdots b_n}$ as $W_{\bar{b}}$ where $\bar{b}$ indicates $b_1 b_2 \ldots b_n$.

Example 5.15 The $W_{\bar{b}}$ representations of the split channels W_2^1 and W_2^2 are illustrated in Fig. 5.3.

Example 5.16 The $W_{\bar{b}} = W_{00}$ representations of the split channel W_4^1 is illustrated in Fig. 5.4.

Example 5.17 The $W_{\bar{b}} = W_{01}$ representations of the split channel W_4^2 is illustrated in Fig. 5.5.

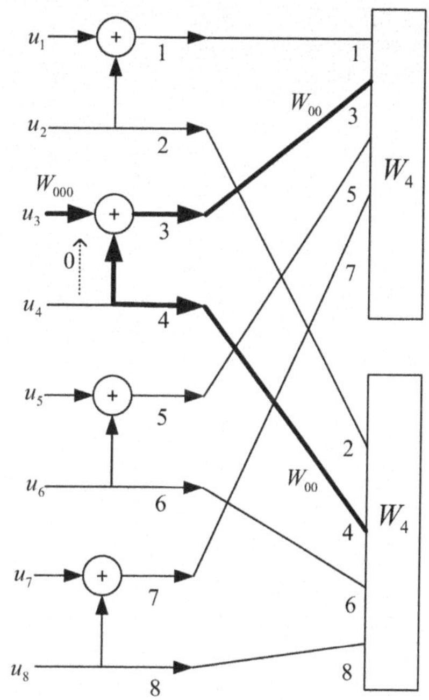

Fig. 5.6 The $W_{\bar{b}} = W_{000}$ representations of the split channel W_4^2

Example 5.18 The $W_{\bar{b}} = W_{000}$ representations of the split channel W_4^2 is illustrated in Fig. 5.6.

Example 5.19 The $W_{\bar{b}} = W_{001}$ representations of the split channel W_8^4 is illustrated in Fig. 5.7.

Let's now state the channel polarization theorem and give its proof.

Theorem 5.1 *Let C be the capacity of any binary discrete memoryless channel W, i.e., B-DMC W. Assume that polar encoding is employed. And let N be the code-word size which is a power of two. As $N \to \infty$, the capacity of some split channels W_N^i approaches to 1, i.e.,*

$$C\left(W_N^i\right) \in (1 - \delta 1] \tag{5.35}$$

and the capacity of some other split channels approaches to 0, i.e.,

$$C\left(W_N^i\right) \in [0\,\delta). \tag{5.36}$$

As $N \to \infty$, the number of split channels with capacity 1 approach to $N \times C$, and the number of split channels with capacity 0 approach to $N - N \times C = N(1 - C)$.

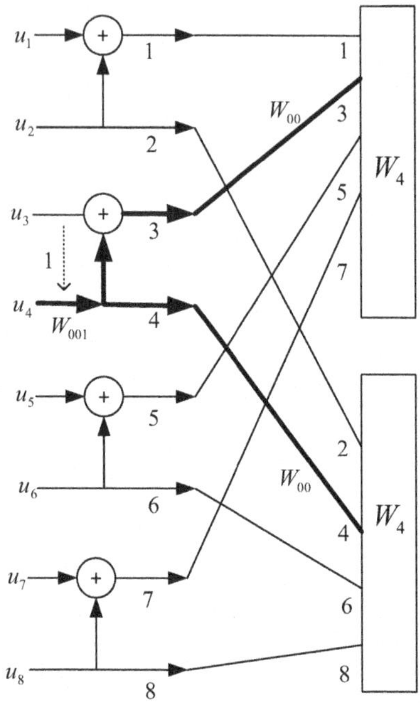

Fig. 5.7 The $W_{\bar{b}} = W_{001}$ representations of the split channel W_8^4

Example 5.20 Let $N = 10000$ be the codeword size of the polar code, and the capacity of a B-DMC be $C = 0.8$. Then, the number of split channels with capacity 1 almost equals to

$$N \times C = 10000 \times 0.8 \to 8000$$

and the number of split channels with capacity 0 almost equals to

$$N(1 - C) = 10000 \times (1 - 0.8) \to 2000.$$

Thus, those 8000 split channels with capacity 1 are chosen for data transmission, and those 2000 split channels with capacity 0 are chosen for frozen bits.

Proof 5.1 We know that if a process is martingale, then it converges to a finite value. For the proof, we will pursue this information. That is, we will show that that the mutual information, or the capacity of the split channels is a martingale stochastic process, and converges either to 0 or 1 as $N \to \infty$. Once we show this, then the proof is complete.

Using the structure in Fig. 5.2, let's first show that the mutual information of split channels is a stochastic process.

Mutual information as a stochastic process:

The random process is nothing but a sequence of random variables defined on the same sample space.

Referring to the structure in Fig. 5.2, we see that the index of the split channel $W_{b_1 b_2 \ldots b_n}$, i.e., the sequence $b_1 b_2 \ldots b_n$ can be considered as being generated by the random number sequence

$$\tilde{B}_1 \tilde{B}_2 \ldots \tilde{B}_n \tag{5.37}$$

where each random variable $\tilde{B}_j$, $j = 1, \ldots, n$ has Bernoulli distribution, i.e., $\tilde{B}_j$ is a Bernoulli random variable. The random variable sequence can be represented by a random process, and this also means that the split channel $W_{b_1 b_2 \ldots b_n}$ can be considered as an instant of the random process. Let's indicate the random process $W_{\tilde{B}_1 \tilde{B}_2 \ldots \tilde{B}_n}$ by $\tilde{K}_n$, i.e.,

$$\tilde{K}_n = W_{\tilde{B}_1 \tilde{B}_2 \cdots \tilde{B}_n}. \tag{5.38}$$

Then, considering the generated values, we can write

$$K_n = W_{b_1 b_2 \ldots b_n}. \tag{5.39}$$

The value of K_{n+1} can either be

$$K_{n+1} = W_{b_1 b_2 \ldots b_n 0} \tag{5.40}$$

or be

$$K_{n+1} = W_{b_1 b_2 \ldots b_n 1} \tag{5.41}$$

with probability 1/2 each. Let's now define the random processes

$$\tilde{I}_n = I\left(\tilde{K}_n\right) \text{ and } \tilde{Z}_n = Z\left(\tilde{K}_n\right). \tag{5.42}$$

Considering the binary sequences $b_1 b_2 \ldots b_n$, we can define a probability space

$$(S, F, P) \tag{5.43}$$

where the sample space S contains all the binary sequences $b_1 b_2 \ldots$, and P is the probability measure defined on S as

$$P(b_1 b_2 \ldots b_n) = \frac{1}{2^n}\ n \geq 1. \tag{5.44}$$

Using F, we can have the filtration

$$F_0 \subseteq F_1 \subseteq \cdots F \tag{5.45}$$

where $F_0 = \{\phi, S\}$.

The random variable $\tilde{B}_j$, and the random processes $\tilde{K}_n$, $\tilde{I}_n$, $\tilde{Z}_n$ are measurable with respect to the F_n.

Proposition 5.1 *The random process $\tilde{I}_n$ is a martingale with respect to the σ-filed F_n $n \geq 0$. That is,*

(1) $F_n \subset F_{n+1}$ *and* $\tilde{I}_n$ *is* F_n *measurable.*

(2)

$$E\left(\left|\tilde{I}_n\right|\right) < \infty$$

(3)

$$\tilde{I}_n = E\left(\tilde{I}_{n+1} | F_n\right)$$

The condition (3) implies that $\tilde{I}_n = E(\tilde{I}_{n+1} | \tilde{I}_n \tilde{I}_{n-1} \ldots \tilde{I}_0)$ *or* $\tilde{I}_n = E(\tilde{I}_{n+1} | \tilde{I}_n)$.

In addition, the random variable sequence, i.e., stochastic or random process, $\tilde{I}_n$ *converges as* $n \to \infty$*, i.e.,*

$$\tilde{I}_\infty = \lim_{n \to \infty} \tilde{I}_n$$

such that

$$E\left(\tilde{I}_\infty\right) = I_0.$$

Proof of the proposition 5.1 The identities in (1) and (2) are correct due to construction and the property $0 \leq \tilde{I}_n \leq 1$.

In (3.23), we stated that

$$I\left(W_{2N}^{2i-1}\right) + I\left(W_{2N}^{2i}\right) = 2I\left(W_N^i\right). \tag{5.46}$$

Considering a cylinder set $S_c(b_1, \ldots, b_n)$, and using (5.46), we can write

$$\begin{aligned} E\left(\tilde{I}_{n+1} \middle| \underbrace{S_c(b_1, \ldots, b_n)}_{\tilde{I}_n}\right) &= \frac{1}{2}\tilde{I}\left(W_{b_1 \ldots b_n 0}\right) + \frac{1}{2}\tilde{I}\left(W_{b_1 \ldots b_n 1}\right) \\ &= \tilde{I}_n. \end{aligned} \tag{5.47}$$

Or in a more explicit manner, we can write that

$$E\left(\underbrace{\tilde{I}\left(W_{b_1\cdots b_n b_{n+1}}\right)}_{\tilde{I}_{n+1}}\middle|\underbrace{\tilde{I}\left(W_{b_1\cdots b_n}\right)}_{\tilde{I}_n}\right)=\frac{1}{2}\tilde{I}\left(W_{b_1\cdots b_n 0}\right)+\frac{1}{2}\tilde{I}\left(W_{b_1\cdots b_n 1}\right)$$
$$=\tilde{I}\left(W_{b_1\cdots b_n}\right)=\tilde{I}_n \tag{5.48}$$

which means that $\{\tilde{I}_n, F_n\}$ is a Martingale.

Thus, if $\{\tilde{I}_n, F_n\}$ is a Martingale, then $\tilde{I}_n$ converges to $\tilde{I}_\infty$. Now, we will show that $\tilde{I}_\infty$ equals to 0 or 1 a.e.

To prove that $\tilde{I}_\infty$ takes value 0 or 1 a.e., we will show in the following proposition that $\tilde{Z}_\infty$ equals to 0 or 1 a.e.

Proposition 5.2 *The random process $\tilde{Z}_n$, i.e., the random variable sequence $\tilde{Z}_n$, together with the Borel field F_n, $n \geq 0$ is a super-martingale satisfying the properties*

(1)

$$F_n \subset F_{n+1} \text{ and the random process } \tilde{Z}_n \text{ is } F_n \text{ measurable.} \tag{5.49}$$

(2)

$$E\left(\left|\tilde{Z}_n\right|\right) < \infty. \tag{5.50}$$

(3)

$$E\left(\tilde{Z}_{n+1}|F_n\right) \leq \tilde{Z}_n, \text{ i.e., } E\left(\tilde{Z}_{n+1}|\tilde{Z}_n, \tilde{Z}_{n-1}, \ldots, \tilde{Z}_0\right)$$
$$\leq \tilde{Z}_n$$
$$\rightarrow E\left(\tilde{Z}_{n+1}|\tilde{Z}_n\right) \leq \tilde{Z}_n. \tag{5.51}$$

Proof of the proposition 5.2 The correctness of (5.49) and (5.50) comes from the definition directly. To prove (5.51), we can proceed as

$$E\left(\underbrace{\tilde{Z}\left(W_{b_1\cdots b_n b_{n+1}}\right)}_{\tilde{Z}_{n+1}}\middle|\underbrace{\tilde{Z}\left(W_{b_1\cdots b_n}\right)}_{\tilde{Z}_n}\right)+\frac{1}{2}\tilde{Z}\left(W_{b_1\cdots b_n 0}\right)+\frac{1}{2}\tilde{Z}\left(W_{b_1\cdots b_n 1}\right) \tag{5.52}$$

in which employing the inequality

$$\tilde{Z}\left(W_{2N}^{2i-1}\right)+\tilde{Z}\left(W_{2N}^{2i}\right) \leq 2\tilde{Z}\left(W_N^i\right) \tag{5.53}$$

we get

$$E\left(\underbrace{\tilde{Z}(W_{b_1\cdots b_n b_{n+1}})}_{\tilde{Z}_{n+1}}\middle|\underbrace{\tilde{Z}(W_{b_1\cdots b_n})}_{\tilde{Z}_n}\right)=\frac{1}{2}\tilde{Z}(W_{b_1\cdots b_n 0})+\frac{1}{2}\tilde{Z}(W_{b_1\cdots b_n 1})\leq\tilde{Z}(W_{b_1\cdots b_n})$$

which means that

$$\left\{\tilde{Z}_n, F_n\right\}$$

is super-martingale. Then, $\tilde{Z}_n$ converges to $\tilde{Z}_\infty$ as $n \to \infty$ such that

$$E\left(\left|\tilde{Z}_n-\tilde{Z}_\infty\right|\right)\to 0. \tag{5.54}$$

Using (5.54), we can also write that

$$E\left(\left|\tilde{Z}_{n+1}-\tilde{Z}_n\right|\right)\to 0. \tag{5.55}$$

In addition to (5.53), we have

$$\tilde{Z}(W_{2N}^{2i})=\left[\tilde{Z}(W_N^i)\right]^2. \tag{5.56}$$

Since (5.56) is satisfied by half of the split channels, we can state that the equality

$$\tilde{Z}_{n+1}=\tilde{Z}_n^2 \tag{5.57}$$

occurs with probability $1/2$. Then, we can write

$$E\left(\left|\tilde{Z}_{n+1}-\tilde{Z}_n\right|\right)\geq\frac{1}{2}E\left(\tilde{Z}_n^2-\tilde{Z}_n\right)\geq 0 \tag{5.58}$$

which can be rearranged as

$$E\left(\left|\tilde{Z}_{n+1}-\tilde{Z}_n\right|\right)\geq\frac{1}{2}E\left(\tilde{Z}_n\left(1-\tilde{Z}_n\right)\right)\geq 0 \tag{5.59}$$

which implies that

$$E\left(\tilde{Z}_n\left(1-\tilde{Z}_n\right)\right)\to 0$$

leading to

$$E\left(\tilde{Z}_\infty\left(1-\tilde{Z}_\infty\right)\right)=0$$

which means that $\tilde{Z}_\infty$ either equals to 0 or 1 a.e. Since $\tilde{I}_\infty = 1 - \tilde{Z}_\infty$, then we have $\tilde{I}_\infty = 0$ or $\tilde{I}_\infty = 1$ a.e.

Thus, we can state that as $N \to \infty$, the split channel capacities $I(W_N^i)$ cluster around 0 and 1.

Before stating Theorem 5.2 let's solve some problems to prepare ourselves for the proof of the theorem.

Example 5.21 Explain the meaning of

$$T_m(\zeta) \triangleq \left\{\bar{b} \in S : Z_i \leq \zeta, \forall i \geq m\right\}. \tag{5.60}$$

where $\bar{b}$ is a binary vector belonging to the cylinder set *S*.

Solution 5.20 $T_m(\zeta)$ simply indicates those split channels with maximum probability of error less than ζ and these split channels have indices greater then *m*, i.e., $i \geq m$.

Example 5.22 Using the expressions

$$Z_{i+1} = Z_i^2 \text{ if } b_{i+1} = 1 \tag{5.61}$$

$$Z_{i+1} \leq 2Z_i - Z_i^2 \leq 2Z_i \text{if } b_{i+1} = 0 \tag{5.62}$$

$$T_m(\zeta) \triangleq \left\{\bar{b} \in S : \ Z_i \leq \zeta, \quad \forall\, i \geq m\right\}.$$

show that for $\bar{b} \in T_m(\zeta)$ and $i \geq m$, we can obtain the inequality

$$\frac{Z_{i+1}}{Z_i} \leq \begin{cases} 2 & \text{if } b_{i+1} = 0 \\ \zeta & \text{if } b_{i+1} = 1. \end{cases} \tag{5.63}$$

Solution 5.21 From (5.62), we get

$$\frac{Z_{i+1}}{Z_i} \leq 2 \quad \text{if } b_{i+1} = 0$$

and from (5.61), we get

$$\frac{Z_{i+1}}{Z_i} = Z_i$$

in which employing $Z_i \leq \zeta$ included in the definition of $T_m(\zeta)$, we obtain

$$\frac{Z_{i+1}}{Z_i} \leq \zeta \quad \text{if } b_{i+1} = 1.$$

Example 5.23 Explain the meaning of

$$U_{m,n}(\eta) \triangleq \left\{ \bar{b} \in S : \sum_{i=m+1}^{n} b_i > \left(\frac{1}{2} - \eta\right)(n-m) \right\} \tag{5.64}$$

where $n > m > 0$ and $0 < \eta < 1/2$.

Solution 5.22 $U_{m,n}(\eta)$ indicates those binary vectors whose partial Hamming weights, i.e., $i \geq m+1$, are larger than $\left(\frac{1}{2} - \eta\right)(n-m)$ where $n > m > 0$ and $0 < \eta < 1/2$.

Theorem 5.2 *For any B-DMC with capacity $I(W) \geq 0$, and with rate R such that $R < I(W)$, there exist a code with a code book A_N, $N = 2^n$, such that $|A_N| \geq NR$ and $Z(W_N^i) \leq O(N^{-5/4})$, $\forall i \in A_N$, i.e., there exist a number of split channels with capacity approaching to 1.*

Proof 5.2 For the recursive split channels, we have the identities

$$Z_{i+1} = Z_i^2 \quad \text{if } b_{i+1} = 1 \tag{5.65}$$

$$Z_{i+1} \leq 2Z_i - Z_i^2 \leq 2Z_i \quad \text{if } b_{i+1} = 0 \tag{5.66}$$

Using (5.65) and (5.66), let's define

$$T_m(\zeta) \triangleq \left\{ \bar{b} \in S : Z_i \leq \zeta, \quad \forall i \geq m \right\}. \tag{5.67}$$

For $\bar{b} \in T_m(\zeta)$ and $i \geq m$, using the identities (5.65) and (5.66), we can obtain the inequality

$$\frac{Z_{i+1}}{Z_i} \leq \begin{cases} 2 & \text{if } b_{i+1} = 0 \\ \zeta & \text{if } b_{i+1} = 1 \end{cases} \tag{5.68}$$

from which, we can write the inequality

$$Z_n \leq \zeta \cdot 2^{n-m} \prod_{i=m+1}^{n} \left(\frac{\zeta}{2}\right)^{b_i}, \bar{b} \in T_m(\zeta) \quad \text{and} \quad n \geq m \tag{5.69}$$

Let's define

$$U_{m,n}(\eta) \triangleq \left\{ \bar{b} \in S : \sum_{i=m+1}^{n} b_i > \left(\frac{1}{2} - \eta\right)(n-m) \right\} \tag{5.70}$$

where $n > m > 0$ and $0 < \eta < 1/2$.

Using (5.70), we can write (5.69) as

$$Z_n \leq \zeta \cdot \left[2^{\frac{1}{2}+\eta}\zeta^{\frac{1}{2}-\eta}\right]^{n-m} \bar{b} \in T_m(\zeta) \cap U_{m,n}(\eta) \tag{5.71}$$

which can be evaluated using the initial values $\zeta_0 = 2^{-4}$ and $\eta_0 = 1/20$ as

$$Z_n \leq 2^{-4-\frac{5(n-m)}{4}} \text{ such that } \bar{b} \in T_m(\zeta_0) \cap U_{m,n}(\eta_0). \tag{5.72}$$

In the sequel, we will show that (5.72) occurs with high probability. It can be shown that [1]

$$Prob\left[T_{m_0}(\zeta)\right] > I_0 - \delta/2 \tag{5.73}$$

where m_0 is a finite integer and $\zeta > 0, \delta > 0$. In addition, using the Chernoff's bound [1] it can be shown that

$$Prob\left[U_{m,n}(\eta)\right] \geq 1 - 2^{-(n-m)[1-H(\frac{1}{2}-\eta)]} \tag{5.74}$$

where H is the binary entropy function. It is possible to find the smallest value of n, i.e., n_0, such that the right hand side of (5.74) satisfies

$$Prob\left[U_{m,n0}(\eta)\right] \geq 1 - \delta/2 \tag{5.75}$$

where the value of n_0 is finite for any $m \geq 0, 0 < \eta < 1/2$, and $\delta > 0$.

Then, it is possible to write the bound

$$P\left[T_{m_1}(\zeta_0) \cap U_{m_1,n}(\eta_0)\right] \geq I_0 - \delta, n \geq n_1$$

where m_1 is the smallest value of m for given ζ_0 and δ values, and n_1 is the smallest value of n for given m_1, η_0, and δ values.

Let's define

$$c \triangleq 2^{-4+\frac{5m_1}{4}}$$

and

$$V_n \triangleq \left\{\bar{b} \in S : Z_n \leq c2^{-\frac{5n}{4}}\right\} n \geq 0 \tag{5.76}$$

then it can be shown that

$$T_{m_1}(\zeta_0) \cap U_{m_1,n}(\eta_0) \subset V_n, \quad n > n_1$$

from which, we get

$$Prob(V_n) \geq I_0 - \delta \quad \text{for}\, n \geq n_1. \tag{5.77}$$

In addition, using (5.76), we can write that

$$\begin{aligned} Prob(V_n) &= \sum_{b_1^n \in X^n} \frac{1}{2^n} 1\left\{Z\left(W_{b_1^n}\right) \leq c2^{-\frac{5n}{4}}\right\} \\ &= \frac{1}{N}|A_N| \end{aligned} \tag{5.78}$$

where $|A_N|$ is defined as

$$|A_N| \triangleq \left\{i : Z\left(W_N^i\right) \leq cN^{-\frac{5}{4}} \text{ and } 1 \leq i \leq N\right\} \tag{5.79}$$

where $N = 2^n$. Using (5.77) and (5.78), we get

$$|A_N| \geq N(I_0 - \delta) \quad \text{for}\, n \geq n_1. \tag{5.80}$$

5.5 Performance Analysis of Polar Codes

For a binary input discrete memoryless channel denoted by W, the maximum probability of transmission error is upper bounded by the Bhattacharyya parameter defined as

$$Z(W) = \sum_y \sqrt{p(y|x=0)p(y|x=1)}$$

which can also be written as

$$Z(W) = \sum_y \sqrt{W(y|0)W(y|1)}$$

where y is the symbol at the output of the channel and 0 and 1 are possible channel inputs.

Proposition 5.3 *If we denote the error event by E, then the probability of E is upper bounded by $Z(W)$, i.e., we have*

$$Prob(\text{E}) \leq Z(W). \tag{5.81}$$

Now let's show the correctness of (5.81) Let x be the channel input and y be the channel output and $\mathcal{X} = \{0, 1\}$, $\mathcal{Y} = \{0, 1, 2, \ldots, M\}$ such that $(x, y) \in \mathcal{X} \times \mathcal{Y}$. We can define the error event as in (5.82)

$$\mathrm{E} = \left\{(x, y) \in \mathcal{X} \times \mathcal{Y} : p\left(x', y\right) \geq p(x, y)\right\} \tag{5.82}$$

which implies that if $x = 0$, then we have

$$p(1, y) \geq p(0, y)$$

or if $x = 1$, then we have

$$p(0, y) \geq p(1, y).$$

The error event in (5.82) can also be expressed as

$$\mathrm{E} = \left\{(x, y) \in \mathcal{X} \times \mathcal{Y} : \hat{x} \neq x\right\}$$

where $\hat{x}$ is the decision result the of input estimator which uses channel output for estimation operation.

The complement expression in (5.82) can be written using the XOR function as $x' = x \oplus 1$, then (5.82) happens to be as (5.83)

$$\mathrm{E} = \{(x, y) \in \mathcal{X} \times \mathcal{Y} : p(x \oplus 1, y) \geq p(x, y)\} \tag{5.83}$$

which can also be written either as

$$\mathrm{E} = \left\{(x, y) \in \mathcal{X} \times \mathcal{Y} : \frac{p(x \oplus 1, y)}{p(x, y)} \geq 1\right\} \tag{5.84}$$

or using a different notation as

$$\mathrm{E} = \left\{(x, y) \in \mathcal{X} \times \mathcal{Y} : \frac{W(x \oplus 1, y)}{W(x, y)} \geq 1\right\}. \tag{5.85}$$

Note that the event E in (5.84) is the set consisting of the pairs (x, y) satisfying

$$\frac{p(x \oplus 1, y)}{p(x, y)} \geq 1.$$

Let's define the indicator function $\mathrm{I}(\cdot)$ for the event E as

$$\mathrm{I}_{\mathrm{E}}(x, y) = \begin{cases} 1 \text{ if} & (x, y) \in \mathrm{E} \\ 0 \; otherwise. \end{cases} \tag{5.86}$$

In (5.84), we have

$$\frac{p(x \oplus 1, y)}{p(x, y)} \geq 1$$

from which we can write

$$\sqrt{\frac{p(x \oplus 1, y)}{p(x, y)}} \geq 1. \tag{5.87}$$

For instance, for $1.2 > 1$ we can write $\sqrt{1.2} = 1.09 > 1$.
Considering (5.86) and (5.87), we get

$$\mathrm{I}_{\mathrm{E}}(x, y) \leq \sqrt{\frac{p(x \oplus 1, y)}{p(x, y)}} \tag{5.88}$$

which can also be written as

$$\mathrm{I}_{\mathrm{E}}(x, y) \leq \sqrt{\frac{W(x \oplus 1, y)}{W(x, y)}} \tag{5.89}$$

The probability of the event E can be calculated using

$$Prob(\mathrm{E}) = \sum_{y} \sum_{x} p(x, y) \mathrm{I}_{\mathrm{E}}(x, y) \tag{5.90}$$

where $\mathrm{I}_{\mathrm{E}}(x, y)$ assures that only those pairs $(x, y) \in \mathrm{E}$ should contribute to the summation. The joint probability expression $p(x, y)$ in (5.90) using $p(x) = 1/2$ can be written as

$$p(x, y) = p(y|x)p(x) \rightarrow p(x, y) = \frac{1}{2} p(y|x).$$

Then, (5.90) happens to be as in

$$Prob(\mathrm{E}) = \sum_{y} \sum_{x} \frac{1}{2} p(y|x) \mathrm{I}_{\mathrm{E}}(x, y) \tag{5.91}$$

in which, employing the inequality (5.88), we get

$$Prob(\mathrm{E}) \leq \sum_{y} \sum_{x} \frac{1}{2} p(y|x) \sqrt{\frac{p(x \oplus 1, y)}{p(x, y)}} \tag{5.92}$$

in which substituting the equalities

$$p(x \oplus 1, y) = p(y|x \oplus 1) \underbrace{p(x \oplus 1)}_{=1/2} \quad p(x, y) = p(y|x) \underbrace{p(x)}_{=1/2}$$

we obtain

$$Prob(\mathrm{E}) \le \sum_{y}\sum_{x}\frac{1}{2}p(y|x)\sqrt{\frac{p(y|x\oplus 1)}{p(y|x)}} \tag{5.93}$$

leading to

$$Prob(\mathrm{E}) \le \sum_{y}\sum_{x}\frac{1}{2}\sqrt{p(y|x)p(y|x\oplus 1)}. \tag{5.94}$$

When (5.94) is expanded for $x = 0$ and $x = 1$, we get

$$Prob(\mathrm{E}) \le \sum_{y}\frac{1}{2}\sqrt{p(y|0)p(y|1)} + \sum_{y}\frac{1}{2}\sqrt{p(y|1)p(y|0)}$$

leading to

$$Prob(\mathrm{E}) \le \sum_{y}\sqrt{p(y|0)p(y|1)}$$

which can also be written as

$$Prob(\mathrm{E}) \le \sum_{y}\sqrt{W(y|0)W(y|1)}$$

resulting in

$$Prob(\mathrm{E}) \le Z(W). \tag{5.95}$$

Thus, we showed that $Z(W)$ can be considered as an upper bound for the maximum probability of error. Let $\mathrm{E_i}$ be the error event for the ith split channel W_N^i, i.e., $\mathrm{E_i}$ denotes the event that the ith bit is decoded incorrectly. Using (5.95), we can write that

$$Prob(\mathrm{E_i}) \le Z(W_N^i).$$

Exercise: Consider the polar encoder for $N = 2$, and assume that binary erasure channels with erasure probabilities $\alpha = 0.5$ are employed. Calculate $Z(W_2^1)$ and $Z(W_2^2)$.

Exercise: Consider the polar encoder for $N = 4$, and assume that binary erasure channels with erasure probabilities $\alpha = 0.5$ are employed. Calculate $Z(W_4^i)$ for $i = 1, 2, 3, 4$.

Block Error Probability:

We will indicate a polar code with the parameter set (N, K, A_u, A_f) where N is the code-word length, K is the number of data bits, A_u is the set of data vectors, and A_f is the set of frozen bit vectors. Let u_1^N, $\hat{u}_1^N$, y_1^N be the information, decoded information, and the received signal sequences respectively. Considering a fixed frozen bit vector employed in code-words, the block error probability for the polar code with the parameters (N, K, A_u, A_f) is calculated as

$$P_{be} = \sum_{u_1^K} \sum_{y_1^N : \hat{u}_1^N \neq u_1^N} p(u_1^N, y_1^N)$$

which can be written as

$$P_{be} = \sum_{u_1^K} \sum_{y_1^N : \hat{u}_1^N \neq u_1^N} p(y_1^N | u_1^N) p(u_1^N)$$

in which substituting

$$p(u_1^N) = \frac{1}{2^K}$$

we get

$$P_{be} = \sum_{u_1^K} \frac{1}{2^K} \sum_{y_1^N : \hat{u}_1^N \neq u_1^N} p(y_1^N | u_1^N)$$

which can also be written as

$$P_{be} = \sum_{u_1^K} \frac{1}{2^K} \sum_{y_1^N : \hat{u}_1^N \neq u_1^N} W_N(y_1^N | u_1^N). \tag{5.96}$$

Considering all the other choices of frozen bit vectors, using (5.96) the average block error probability can be calculated as

$$P_{avg} = \sum_{\bar{u}_f \in A_c} \frac{1}{2^{N-K}} P_{be}.$$

Proposition 5.4 *For any discrete memoryless channel, the polar code with the parameters (N, K, A_u) has an average block probability of error P_{avg} upper bounded as in*

$$P_{avg} \leq \sum_i Z(W_N^i).$$

And for any specific choice of frozen vector $\bar{u}_f$ the block error probability P_{be} is upper bounded as in

$$P_{be} \le \sum_i Z(W_N^i).$$

Proof 5.4 Let $\mathcal{X}_1^N$ and $\mathcal{Y}_1^N$ be the word, i.e., vector, sets for the code and received words respectively. Let's define the event $\mathrm{B_i}$ as

$$\mathrm{B_i} = \left\{ \left(u_1^N, y_1^N\right) \in \mathcal{X}_1^N \times \mathcal{Y}_1^N : \hat{u}_1^{i-1} \neq u_1^i, \; \hat{u}_i \neq u_i \right\}$$

which indicates the event where the first decision error occurs at bit location i. That, is the event $\mathrm{B_i}$ consists of the pairs (u_1^N, y_1^N) such that the first decision error occurs at bit index i, before i, all the bits are received correctly.

Considering $\mathrm{B_i}$, the block error event can be defined as

$$\mathrm{E} = \bigcup_i \mathrm{B_i}.$$

Now, let's consider the split channel W_N^i. Following a similar reasoning as in the derivation of (5.95), we can define

$$\mathrm{E_i} = \left\{ \left(u_1^N, y_1^N\right) \in \mathcal{X}_1^N \times \mathcal{Y}_1^N : \frac{W_N^{i-1}\left(y_1^N, u_1^{i-1} | u_i \oplus 1\right)}{W_N^{i-1}\left(y_1^N, u_1^{i-1} | u_i\right)} \ge 1 \right\}. \tag{5.97}$$

The event $\mathrm{E_i}$ involves the pairs (u_1^N, y_1^N) such that at location i decision error occurs without concerning the decision of the previously decoded bits. Due to this reasoning, we can write that

$$\mathrm{E} \subset \bigcup_i \mathrm{E_i}$$

from which we obtain

$$Prob(\mathrm{E}) \le \sum_{\mathrm{i}} Prob(\mathrm{E_i}).$$

Now let's consider the calculation of $Prob(\mathrm{E_i})$. Considering (5.97), and following a similar approach as in derivation of (5.89) and (5.90), we can write $Prob(\mathrm{E_i})$ as in

$$Prob(\mathrm{E_i}) = \sum_{u_1^i, y_1^N} \sqrt{\frac{Prob\left(y_1^N, u_1^{i-1} | u_i \oplus 1\right)}{Prob\left(y_1^N, u_1^{i-1} | u_i\right)}} Prob\left(y_1^N, u_1^i\right)$$

which can be shown to be equal to

$$Prob(\mathrm{E_i}) = Z\left(W_N^i\right). \tag{5.98}$$

In fact, the calculation of (5.98) is explained in details in the following sub-section. Thus, we can write the upper bound expression for block error probability as

$$Prob(\mathrm{E}) \leq \sum_{\mathrm{i}} Z\left(W_N^i\right).$$

Proposition 5.5 *The Bhattacharyya parameter of the split channel W_N^i can be considered as the expected value of the squared root of the likelihood ratio random variable, and it is defined as*

$$Z\left(W_N^i\right) = E\left\{\sqrt{\frac{W_N^i\left(\tilde{Y}_1^N, \tilde{U}_1^{i-1}|\tilde{U}_i \oplus 1\right)}{W_N^i\left(\tilde{Y}_1^N, \tilde{U}_1^{i-1}|\tilde{U}_i\right)}}\right\}. \tag{5.99}$$

Proof 5.5 The mathematical expression

$$\sqrt{\frac{W_N^i\left(\tilde{Y}_1^N, \tilde{U}_1^{i-1}|\tilde{U}_i \oplus 1\right)}{W_N^i\left(\tilde{Y}_1^N, \tilde{U}_1^{i-1}|\tilde{U}_i\right)}}$$

can be considered as a function of $\tilde{U}_1^i, \tilde{Y}_1^N$, i.e.,

$$g\left(\tilde{U}_1^i, \tilde{Y}_1^N\right) = \sqrt{\frac{Prob\left(\tilde{Y}_1^N, \tilde{U}_1^{i-1}|\tilde{U}_i \oplus 1\right)}{Prob\left(\tilde{Y}_1^N, \tilde{U}_1^{i-1}|\tilde{U}_i\right)}}.$$

Then, the expected value of $g\left(\tilde{U}_1^i, \tilde{Y}_1^N\right)$ can be calculated using

$$E\left(g\left(\tilde{U}_1^i, \tilde{Y}_1^N\right)\right) = \sum_{u_1^i, y_1^N} g\left(u_1^i, y_1^N\right) p\left(u_1^i, y_1^N\right) \tag{5.100}$$

where $p(u_1^i, y_1^N)$ is the joint probability mass function between u_1^i and y_1^N, and it can be written as

$$p\left(u_1^i, y_1^N\right) = p\left(y_1^N, u_1^{i-1}|u_i\right) \underbrace{p(u_i)}_{=1/2} \rightarrow p\left(u_1^i, y_1^N\right) = \frac{1}{2} p\left(y_1^N, u_1^{i-1}|u_i\right). \tag{5.101}$$

If (5.101) is used in (5.100), we obtain

$$E\left(g\left(\tilde{U}_1^i, \tilde{Y}_1^N\right)\right) = \frac{1}{2} \sum_{u_1^i, y_1^N} \sqrt{\frac{Prob\left(y_1^N, u_1^{i-1}|u_i \oplus 1\right)}{Prob\left(y_1^N, u_1^{i-1}|u_i\right)}} Prob\left(y_1^N, u_1^{i-1}|u_i\right)$$

which can also be written as

$$E\left(g\left(\tilde{U}_1^i, \tilde{Y}_1^N\right)\right) = \frac{1}{2}\sum_{u_1^i, y_1^N}\sqrt{\frac{W_N^i\left(y_1^N, u_1^{i-1}|u_i \oplus 1\right)}{W_N^i\left(y_1^N, u_1^{i-1}|u_i\right)}}W_N^i\left(y_1^N, u_1^{i-1}|u_i\right) \quad (5.102)$$

which can be simplified as

$$E\left(g\left(\tilde{U}_1^i, \tilde{Y}_1^N\right)\right) = \frac{1}{2}\sum_{u_1^i, y_1^N}\sqrt{W_N^i\left(y_1^N, u_1^{i-1}|u_i \oplus 1\right)W_N^i\left(y_1^N, u_1^{i-1}|u_i\right)}. \quad (5.103)$$

When the summation in (5.103) is expanded for $u_i = 0$ and $u_i = 1$, we obtain

$$E\left(g\left(\tilde{U}_1^i, \tilde{Y}_1^N\right)\right) = \frac{1}{2}\sum_{u_1^i, y_1^N}\sqrt{W_N^i\left(y_1^N, u_1^{i-1}|0 \oplus 1\right)W_N^i\left(y_1^N, u_1^{i-1}|0\right)} + \frac{1}{2}\sum_{u_1^i, y_1^N}\sqrt{W_N^i\left(y_1^N, u_1^{i-1}|1 \oplus 1\right)W_N^i\left(y_1^N, u_1^{i-1}|1\right)}$$

which leads to

$$E\left(g\left(\tilde{U}_1^i, \tilde{Y}_1^N\right)\right) = \sum_{u_1^i, y_1^N}\sqrt{W_N^i\left(y_1^N, u_1^{i-1}|1\right)W_N^i\left(y_1^N, u_1^{i-1}|1\right)}$$

which is nothing but the Bhattacharyya parameter of the split channel W_N^i. Thus, we showed that

$$Z\left(W_N^i\right) = E\left\{\sqrt{\frac{W_N^i\left(\tilde{Y}_1^N, \tilde{U}_1^{i-1}|\tilde{U}_i \oplus 1\right)}{W_N^i\left(\tilde{Y}_1^N, \tilde{U}_1^{i-1}|\tilde{U}_i\right)}}\right\}. \quad (5.104)$$

Problems

(1) Calculate the mean value and variance of Bernoulli random variable.

(2) The random process $\tilde{X}(t)$ is defined as $\tilde{X}(t) = \tilde{A}\cos(wt + \tilde{\theta})$ where $\tilde{A}$ and $\tilde{\theta}$ are independent random variables. Find the mean value $m(t)$ of this random process.

(3) The random process $\tilde{X}(t)$ is defined as $\tilde{X}(t) = \tilde{A}$ where $\tilde{A}$ is a Bernoulli random variable. Find the mean value $m(t)$ of this random process.

(4) Show the path between the input and outputs of the split channels W_4^1, W_4^3, and represent the paths by binary sequences, and calculate the probability of each path.

References

1. E. Arıkan, Channel polarization: a method for constructing capacity-achieving codes for symmetric binary-input memoryless channels. IEEE Trans. Inf. Theory **55**(7), 3051–3073 (2009)
2. A. Andi, O. Gazi, Fast decoding of polar codes, IET Communications, submitted, (2017)
3. E. Arıkan, A performance comparison of polar codes and Reed-Muller codes. IEEE Comm. Lett. **12**(6), June (2008)

O. Gazi, *Polar Codes*, Springer Topics in Signal Processing 15,
https://doi.org/10.1007/978-981-13-0737-9

Index

O. Gazi, *Polar Codes*, Springer Topics in Signal Processing 15,
https://doi.org/10.1007/978-981-13-0737-9

The manufacturer's authorised representative in the EU is Springer Nature Customer Service Centre GmbH, Europaplatz 3, 69115 Heidelberg, Germany. If you have any concerns regarding our products, please contact ProductSafety@springernature.com

Printed and bound by CPI Group (UK) Ltd, Croydon, CR0 4YY
25/09/2026
02222894-0015